ART DECO
新装饰主义室内设计
INTERIOR DESIGN

金盘地产传媒有限公司　策划　　高迪国际出版（香港）有限公司　编著

中国林业出版社
China Forestry Publishing House

图书在版编目（CIP）数据

新装饰主义室内设计/高迪国际出版（香港）有限公司编著. -- 北京：中国林业出版社, 2017.1
ISBN 978-7-5038-8441-2

Ⅰ. ①新… Ⅱ. ①高… Ⅲ. ①室内装饰设计 Ⅳ. ①TU238

中国版本图书馆 CIP 数据核字 (2016) 第 048666 号

新装饰主义室内设计

编　　著	高迪国际出版（香港）有限公司
责任编辑	纪　亮　王思源
策划编辑	邹筠娟
文字编辑	高雪梅
英文编辑	赖小珍、李嘉珍
装帧编辑	高迪设计
封面设计	杨丽冰
出版发行	中国林业出版社
出版社地址	北京西城区德内大街刘海胡同 7 号，邮编：100009
出版社网址	http://lycb.forestry.gov.cn/
经　　销	全国新华书店
印　　刷	深圳市汇亿丰印刷科技有限公司
开　　本	300 mm × 300 mm
印　　张	35
版　　次	2017 年 1 月第 1 版
印　　次	2017 年 1 月第 1 次印刷
标准书号	ISBN 978-7-5038-8441-2
定　　价	599.00 元（精）

图书如有印装质量问题，可随时向印刷厂调换（电话：0755-82413509）。

FORWORD
前言

ART DECO（装饰主义）风格起源
——ART DECO（装饰主义）的前世今生

20世纪初，在大工业迅速发展、商业日益繁荣的形势推动下，许多设计师开始寻找一种符合现代生活特征的装饰形式。

1922年，英国考古学家在埃及发现图坦卡蒙墓——这个精妙绝伦的古典艺术世界霎时轰动了欧洲的先进设计师圈子。

那些美艳于3300年前的绝世古物，特别是图坦卡蒙的金面具，仅有简单的几何图形，使用金属色系列和黑白色彩系列，却达到高度装饰的艺术效果，这一切都给予设计师们有力而无限的创作启示，也成为了之后声名鹊起的Art Deco（装饰主义）装饰艺术风格最实用的创意源泉之一。

1925年法国的"世界艺术装饰和工业博览会"的开幕更象征了"装饰主义"Art Deco（装饰主义）风格的成型。从此Art Deco（装饰主义）快速成为影响西方世界的全新艺术风格。

Art Deco（装饰主义）风格从它发轫和诞生的那一刻起，将近百年来，始终有一批狂热的推广者和追随者沉醉在这种风格所营造的优雅、永不落伍的摩登氛围中。你再跑到街上看看吧，Art Deco（装饰主义）无处不在，在历经多次改天换地的运动之后，它仍然骄傲地挺立在大街小巷的某处建筑或室内设计之中。

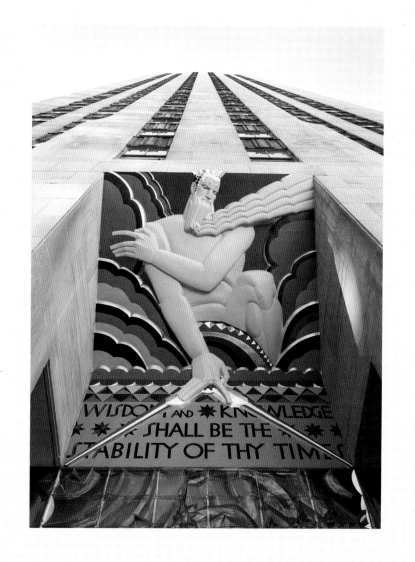

ART DECO（装饰主义）风格发展
——走进 ART DECO（装饰主义）的风华年代

从 20 世纪 30 年代起，Art Deco（装饰主义）风格建筑渐渐成为了那个时代的精神图腾，一直到现在，Art Deco（装饰主义）从一种经久不衰的建筑风格演变成了一种生活态度和生活方式，被很好地融入到室内装饰中去。

随着欧美帝国资本主义向外扩张，远东、中东、埃及、罗马、希腊及马雅等古老的文化的物品或者图腾，它们成了 Art Deco（装饰主义）风格的素材来源。而 Art Deco（装饰主义）风格也是艺术装饰风格，它发源于法国，兴盛于美国，是世界建筑史上的一个重要的风格流派。

Art Deco（装饰主义）在 30 年代让现代主义与装饰性幻想的共存，似乎是一种矛盾的倾向。踢踏舞、Jazz 和大工厂，为人们带来越洋班机和摩天大楼，也渲染了一个歌舞升平的好莱坞时代。它反映了人们当时心中理想的生活模式。

Art Deco（装饰主义），打破了现代主义的冰冷面具，让建筑真正从工业产品中脱离出来，补加以人的气息。30 年代的设计师们在营造一种俏皮、欢快的氛围。比如迈阿密的沙滩旅馆，粗重的窗楣，多变的色彩，无不诉说着一种欢乐、向上的进步气息，人体的曲线、动感的线条被大量使用。新的技术新的材料并没有被建筑师所摒弃，只是不再机械、冷酷的站立在城市中，并从各种风格文化中广泛地吸收营养。

ART DECO（装饰主义）在中国

大上海的 Art Deco（装饰主义）情缘

早在上海开埠之初，中国传统的建筑样式就迅速被西方的建筑风格所取代。在 20 世纪 20 年代末以前，上海基本都是西方复古主义建筑风格的天下，诸如汇丰银行大楼、大华饭店、上海邮政总局等建筑，被誉为"完全能与欧美一流的学院派复古建筑媲美"的作品。

但随着20世纪二三十年代西方世界Art Deco（装饰主义）风潮的来袭，灵活善变的大上海也不甘落后地跟上了浪潮的步伐。从20世纪20年代后半段开始，上海的建筑已经明显转向了Art Deco（装饰主义）风格。1928年建成的德义公寓可以说是上海最早建成的Art Deco（装饰主义）建筑，其建筑立面上精美的V形和卷涡状几何石雕正是当时最摩登的Art Deco（装饰主义）装饰元素。而此后一年，于1929年9月落成的沙逊大厦则是上海建筑全面走向装饰艺术派时期的标志。

自己心中的现代建筑样式。

譬如1929年，由杨锡设计的南京饭店就是上海已知的最早由中国建筑师设计建成的Art Deco（装饰主义）风格的建筑。而两年之后同样由他设计的百乐门舞厅已经拥有非常娴熟的Art Deco（装饰主义）表现技法。但只是一味模仿欧美建筑并不能让中国建筑师满足。正如一切外来文化都必然要经历一场"本土化"运动一样，随着中国建筑师队伍的成长，Art Deco（装饰主义）在传入上海不久也开始与有地方特色的装饰题材相结合，并吸收了一些中国的古典装饰元素。

到了20世纪40年代，随着人们对这种新兴的装饰风格的新鲜感逐步消失和此种装饰艺术主要服务于上流社会的受众限制，Art Deco（装饰主义）风格开始走向衰落，并迅速淡出了人们的视野。而一贯紧跟欧美潮流的上海也毫不意外地将这种曾经风靡一时的艺术风格抛到了身后。虽然在20世纪的后半段，Art Deco（装饰主义）的传奇仍不断被后人以各种方式纪念、效仿、复兴与再现，但属于它的时代毕竟已经成为历史。

Art Deco（装饰主义）在天津

20世纪20年代，Art Deco（装饰主义）在天津也历经繁华和沧桑。天津的五大道会聚着大量Art Deco（装饰主义）风格的建筑，这让天津成为中国北方唯一拥有Art Deco（装饰主义）建筑群的城市。2010年重新装修后的天津利顺德大饭店，恢复了那个年代Art Deco（装饰主义）的装饰风貌。天津的五大道会聚着大量Art Deco（装饰主义）风格的建筑，这让天津成为中国北方唯一拥有Art Deco（装饰主义）建筑群的城市。

入乡随俗的Art Deco（装饰主义）风格

与此同时，一批曾留学欧美的中国建筑师也主动摆脱掉学院派教育的思想桎梏，用Art Deco（装饰主义）这种流行符号来构筑

21世纪中国的"Art Deco（装饰主义）复兴"

当国外的Art Deco（装饰主义）建筑已成历史建筑，而中国各地城市的Art Deco（装饰主义）建筑，尤其是Art Deco（装饰主义）住宅建筑方兴未艾，被视作传世经典而大放异彩。

21世纪以来，中国各地Art Deco（装饰主义）建筑的建设规模，计有上百个楼盘数百上千万平方米，超过了历史上世界Art Deco（装饰主义）建筑的总量。不能不说中国在21世纪初的建筑时代中，掀起了一场空前的"Art Deco（装饰主义）复兴"。

按照Art Deco（装饰主义）住宅建筑在中国各城市出现的年代排序，上海在2005年，杭州、成都在2006年，天津在2007年，广州、太原2009年，都先后引进了Art Deco（装饰主义）风格建筑。2010年，深圳出现首个Art Deco（装饰主义）建筑风格的楼盘。

ART DECO（装饰主义）风格特点

一、标志性特点：简洁又不失装饰性

Art Deco（装饰主义）风格兼具古代的（古埃及、玛雅、阿兹特克文化等）、古典（装修效果图）的、哥特的、立体派的种种艺术元素和手法，同时又结合了因工业文化所兴起的机械美学和爵士时代的摩登表现手法，所以它所包含的艺术元素是十分多样的。

标志性的装饰特点有：阶梯状收缩造型（20世纪的象征）、放射状线形太阳光与喷泉造型（新时代曙光的象征）、几何图形（机械与科技的象征）、全新题材的摩登雕塑和浅浮雕与建筑立面的结合（当时社会对科技文明的向往）、新女性的形体（女性解放与女权的象征）、古老文化的纹样（对埃及与中美洲古老文明的想象）、速度、力量与飞行的流线造型（交通运输飞速发展的象征）、大胆艳丽的色彩、对新材料的使用和对名贵材料的偏好等。

缤纷、华丽的装饰图案，亦饰以装饰艺术派的图案和纹样，比如麦穗、太阳图腾等等，显现出华贵的气息。

三、线条特点：对称、重复及渐变形式美学

在室内设计中，Art Deco（装饰主义）所体现出来的是基于线条形式的强烈的装饰性，以及在原则上通过灵活地运用对称、重复、渐变等形式美学的法则。从而使简单的几何造型充满戏剧性，在最终呈现的形式造型上富于一定的装饰性。

回纹饰曲线线条、金字塔造型等埃及元素纷纷出现在室内设计上，表达了当时高端阶层所追求的高贵感，而摩登的形体又赋予古老的、贵族的气，代表的是一种复兴的城市精神。通过新颖的造型、艳丽夺目的色彩以及豪华材料的运用，成为一种摩登艺术的符号。

二、造型特点：机械式的完美主张

Art Deco（装饰主义）拥有机器时代的技术美感，汽车的造型、技术性能、速度感成为设计的推动力。汽车机械零件的直线，棱角，齿轮组的简洁明快，都成为设计师的灵感来源。机械式、几何图案、纯综装饰成为它标志性的特点，典型的装饰图案如扇形辐射状的太阳光、齿轮或流线型线条及对称简洁的几何构图等等。

Art Deco（装饰主义）风格，常用方形、菱形和三角形作为形式基础，运用于地毯、地板、家具贴面等处，创造出许多繁复、

四、色彩特点：塑造情感的重要方式

Art Deco（装饰主义）之所以备受人们推崇和喜爱，是因为它本身有一种力量，除了富有几何抽象造型以外，能够给人带来强烈的视觉冲击的，还有就是它的五彩斑斓和激烈昂扬的色彩塑造，而这种冲击很大程度来源于直观的色彩。Art Deco（装饰主义）在色彩构成上与新艺术运动和工业美术运动追求典雅色彩大相径庭，它特别强调的是强烈的纯色、对比色和经典的金属色系，比如亮丽的红色、吓人的粉红色、电器类的蓝色、警报器的黄色，到探戈的橘色、及带有金属味的金色、银白色以及古铜色等等。

各种不搭调的色彩混搭在一起，让空间多姿多彩，充满创意，给人造成华美绚烂的视觉印象。同时色彩的组合，能让这些饱和的色彩为人们带来激烈昂扬的情绪。时色彩的组合，能让这些饱和的色彩为人们带来激烈昂扬的情绪。

窗帘、沙发坐靠面都采用了热情的枣红色，跟稳重的咖啡色混搭在一起，色彩的浓郁让它们醒目，与此同时又给它们增加了一层神秘的面纱。在现今强调个性和张扬独立精神的时代特征下，色彩理所当然也就成为寄托精神和表达情感的重要工具了，室内设计也不再是一如既往的白色，在此空间中热烈的红色、深沉的黑色以及各种华丽的壁纸使空间多姿多彩，活泼又不失稳重，充满创意。

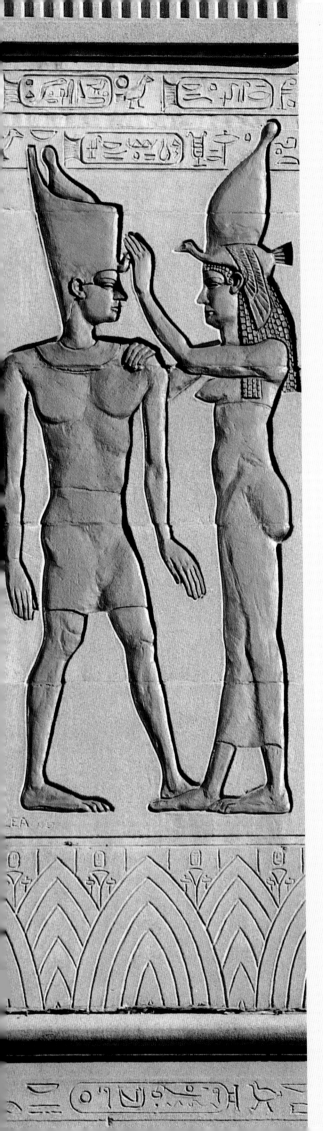

ART DECO（装饰主义）风格室内运用
——开启另类室内装饰艺术时代

一、浓烈又精致的室内装饰品，自然又个性的材质搭配

当前的室内设计师沿用了家具饰面材质、选料精美，使用镶嵌贝壳等特殊工艺，传承欧洲经典，使其显得高贵优雅。有的家具在深沉的黑色贴面材料中镶嵌洁白的贝壳，画龙点睛，流出七彩光晕；有的用胡桃木做饰面，饰以看似平滑而色彩浓重的几何图案；腿脚的螺旋纹理依然传承经典，于张扬中内敛温柔。

同时植物的造型也是Art Deco（装饰主义）风格设计中的常用元素。在这种风格中通常使用大枝叶大线条的干枝干花类，然而不同于新艺术风格对植物纹样的模仿，而是将植物造型抽象升华为几何造型。经过提炼的图案表现出特殊的装饰性，夸张造型中既保留了华丽的风格又具有强烈的现代气息。

在艺术装饰品（如雕塑、壁画等）的选用上，雕塑通常选用富于曲线感的过渡作品，壁画通常选用抽象或富于异国情调的作品。

Art Deco（装饰主义）风格中对金属、玻璃的运用同样已经成了其灵魂所在，通常情况下，此类材质用于建筑内外门窗线脚、檐口及建筑腰线、顶角线等部位，内部门窗、栏杆、家具细部等。

具体来说，家具材料多采用实木，保留材料本身的纹理和色泽并且通过色彩对比，通常是红—黑—红，产生强烈装饰性。局部采用金色和银色点缀于线脚和转折面，强调家具的结构和质感以及雍容华贵的气质。材料本身依然可以通过二次加工产生别样风情，例如表面进行图案刻画，并用抛光金属镶嵌和对比。而地面材料通常运用花岗岩或大理石进行图案拼接和创造，多用例如金线米黄、啡网纹、黑金沙等机理丰富多样的材料来体现华贵气质。除此以外，Art Deco（装饰主义）风格中，对富于异国特征的材质的运用也非常普遍，例如，对中国瓷器、丝绸，非洲木雕，日式锦帛，东南亚棉麻，法国宫廷烛台等，都很好地丰富了Art Deco（装饰主义）的内涵与形式。

有人认为，都市人的怀旧创造了Art Deco（装饰主义）在当今社会的流行和时尚。我们并不否认这样的解释，但不仅仅因为怀旧才能时尚，而是一种审美价值观念的体现以及对传统文化的继承和发扬才造就了时尚。对于经典的简单复制和怀旧并不是好的设计，时尚的生命力在于从经典中寻找精髓和品质，在现代审美的理解基础上的传承与创新。

二、Art Deco（装饰主义）室内设计与软装技巧

（1）Art Deco（装饰主义）勇于尝试诸如钢铁、玻璃等新材料，并运用一些豪华的装饰来提升设计品位，比如青铜、名贵的纺织品，比较注重表现材料的质感、光泽。

（2）Art Deco（装饰主义）造型设计中多采用几何形状或用折线进行装饰；线条明朗，过渡温和，表现夸张，纹饰多种多样，有意大利未来主义的太阳光芒、闪电纹样，而中国

式样的 Art Deco（装饰主义）也以传统花卉例如牡丹入画。

（3）Art Deco（装饰主义）色彩构成上与新艺术运动和工业美术运动追求典雅色彩大相径庭，特别强调强烈的纯色、对比色和金属色系，包括鲜红、鲜黄、鲜蓝，金属色例如古铜、金、银等色彩，造成华美绚烂的视觉印象。

（4）Art Deco（装饰主义）同样彰显着时代的特征，比如以当时时髦的汽车造型作底座的灯饰品，又比如大量运用曲线优美的女性形状。

（5）选择造型夸张的植物点缀，通常配以简单几何感的花瓶，提升空间温情度。

（6）墙纸在如今室内设计中并没有被淘汰，反而由于技术的改进，更方便使用和清洗。特别是一些艺术性很强的墙纸能塑造出雍容华贵的气氛。

（7）在装饰细节，如踢脚、顶角以及柜子和床的线脚使用一些结构上的小小重复能够有效增加家具和空间的精致度。

ART DECO（装饰主义）风格代表大师

Art Deco（装饰主义）的代表人物不在少数，如法国的 E.J. 鲁夫曼和 G.Guevrekian；美国的 F. Steele 和 P.T. 弗兰克尔等。其中，美国建筑大师 Frank Lloyd Wright（古根海姆美术馆和流水别墅（装修效果图）的设计者）对 Art Deco（装饰主义）风格的发展有着特别的贡献。他在 20 世纪早期的设计被认为是美国摩登装饰的萌芽。他最早尝试的是一种自然题材的风格化装饰；同时又受到欧洲当代设计尤其是维也纳分离派的影响，在设计中采用趋于几何形的装饰。他为团结神庙（Unity Temple）做的室内设计就强调直线、正方形和长方形等主题。

之后，他进一步将个人风格与美国本土的文化特点相结合，他借鉴美洲的玛雅和阿兹特克文化，以此来创造属于美国的现代（装修效果图）风格。他设计的芝加哥米德威游乐场（MidwayGarden）和他为东京帝国饭店设计的瓷砖（装修效果图）等都被认为是在 1925 年巴黎博览会以前出现的最早的摩登设计。

他采用的许多装饰题材和设计手法，后来都在美国 Art Deco（装饰主义）建筑、家具和室内设计中被广泛运用。可惜，他的才华一开始并未受到美国人的重视，他在早期对美国的影响都是通过欧洲人的宣扬间接产生的。

CONTENTS
目录

012
FACADE & FRONT
门头门脸

128
FACADE ORNAMENT
立面装饰

204
CEILING
天花板

244
FLOOR PATTERN
地面图案

270
WINDOW
窗户

284
STAIRS
楼梯

308
CLOCK
钟表

322
LIGHTING
灯具

350
WAYFINDING & SIGNAGE
导视标识

366
FURNITURE & FURNISHINGS
家具摆设

374
CORRIDOR SPACE
过道空间

400
RAILING & BARRIER
护栏栏杆

FACADE & FRONT
门头门脸

Art Deco（装饰主义）风格多元化，能够达到时尚与复古的独特结合。注重表现材料的质感、光泽；造型设计中多采用几何形状或用折线进行装饰；色彩设计中强调运用鲜艳的纯色、对比色和金属色，造成华美绚烂的视觉印象。

犹如金属屏风般的门面设计，黑色铁件镂刻出 Art Deco（装饰主义）的经典图腾线条，部分线条边框漆上金色油漆，鲜明强烈的视觉效果。外墙色彩、丰富的纵向直线条装饰和逐层退缩的结构轮廓，都是 Art Deco（装饰主义）风格的体现。

FACADE & FRONT

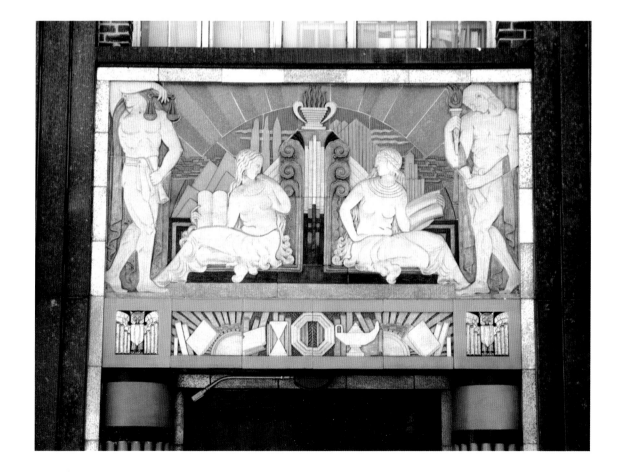

门头门脸

1. Golden and blue elegant decorated door, Les Invalides in Paris.
巴黎荣军院装饰精美的金蓝色大门。

2. Door handel from cupboard from youth style period.
青年风格时期的壁橱上的门把手。

3. Doorway to the past. Art Nouveau door handles on the Corones Hotel in Charleville built between 1924 and 1929.
古时的大门。建于 1924 至 1929 年间的沙勒维尔克洛尼斯酒店的新艺术风格门把手。

4/5/6. Newark Art Deco.
纽瓦克装饰艺术。

FACADE & FRONT

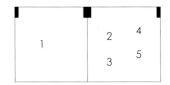

1/2/3/4/5. St. Petersburg Art Deco.
圣彼得堡装饰艺术。

门头门脸

FACADE & FRONT

1. Doors at the Mann Chinese in Hollywood, CA.
加利福尼亚州好莱坞中国剧院大门。

2. Park Ave Station.
公园大道车站。

3. Newark Art Deco.
纽瓦克装饰艺术。

4. Marine Building Vancouver.
温哥华海军大厦。

5. Exterior sea life details of the Marine Building.
海军大厦外部的海上生物特写。

FACADE & FRONT

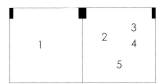

1. Detail of the monument in Vienna, Austria.
奥地利维也纳纪念碑细部。

2/3/4. Paramount Theater.
派拉蒙剧院。

5. Havana Art Deco.
哈瓦那装饰艺术。

门头门脸

FACADE & FRONT

1. Art Deco facade at The Fred F. French Building in Manhattan. The National Register of Historic Places listed the building.
曼哈顿弗拉德·F法国大厦的装饰艺术立面。该建筑被列入《国家历史名胜名录》。

2. Prague Art Deco.
布拉格装饰艺术。

3. Interior of Palazzo Vecchio (Old Palace) with antique frescoes and the fountain with the statue of the putto with dolphin in Florence, Italy.
意大利佛罗伦萨韦奇奥宫内部，装饰有精美的壁画，以及饰有丘比特与海豚雕塑的喷泉。

4. Glise Saint Nicaise.
圣尼凯斯教堂。

门头门脸

FACADE & FRONT

门头门脸

1. The entrance of Arequipa built in 1929.
阿雷基帕入口，建于 1929 年。

2. Marine Building Vancouver.
温哥华海事大厦。

3. Havana.
哈瓦那。

4. Incarnation Episcopal Church Door and Stained Glass Window by Louis Comfort Tiffany, New York City.
纽约路易斯·康福特·蒂凡尼设计的圣公会化身大门和彩色玻璃。

5. Standard Building built in 1924.
标准大厦，建于 1924 年。

FACADE & FRONT

门头门脸

1. Cleveland Art Deco.
克利夫兰装饰艺术。

2. Kogen-Miller Studios.
高原·米勒工作室。

3. Art Deco doorway, West 4th Street at 7th Avenue, Greenwich Village, New York City.
纽约格林威治村第七大道西四街装饰艺术大门。

4. A beautiful decorated art-deco wooden door.
一个装饰精美的装饰艺术木门。

5. Cleveland Art Deco.
克利夫兰装饰艺术。

6. Aldred Building.
阿尔德雷德大厦。

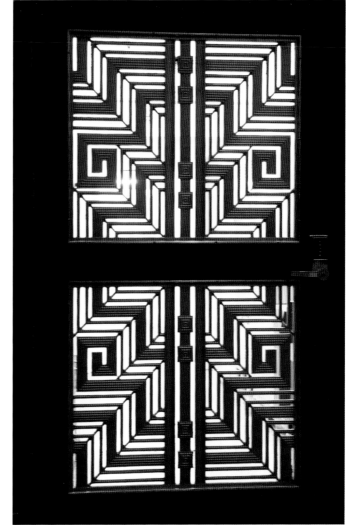

1/2/3. National Breweries.
国家酿酒厂。

4. St Olaf House-St. Olaf House, Hay's Wharf (1929-31) by H.S. Goodhart-Rendel.
干草码头圣奥拉夫楼（1929年至1931年），由斯莱德．古德哈特－林德宪设计。

5. New York Art Deco.
纽约装饰艺术。

6. Newark, Art Deco.
纽瓦克装饰艺术。

1. National Museum of Iceland.
冰岛国家博物馆。

2. Queen Mary Hotel.
玛丽皇后酒店。

3. Aldred Building.
阿尔德雷德大厦。

4/5. Eaton's College Street Store.
伊顿学院街店。

门头门脸

FACADE & FRONT

1. Empire State Building.
帝国大厦。

2. Art deco detail of a door.
大门的装饰艺术细节。

3. Prague Art Deco.
布拉格装饰艺术。

4. Silesian Parliament.
西里西亚议会。

5. Golden decoration of the door to Cathedral of the Dormition in Kiev Pechersk Lavra - famous monastery inscribed on UNESCO world heritage list.
基辅洞窟修道院的安息大教堂的金色装饰大门，被列入联合国教科文组织世界遗产的著名寺院。

6. NBC Tower.
美国全国广播公司大厦。

门头门脸

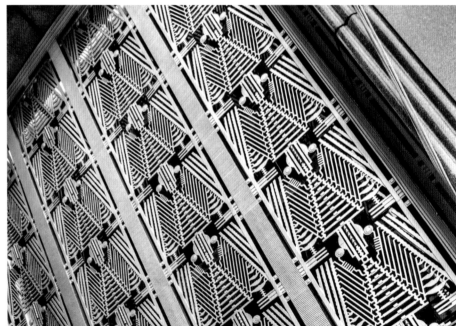

1/2/3. 120 Wall Street is a skyscraper in Wall Street, New York City, United States which was completed in 1930.
华尔街 120 号，是美国纽约华尔街的一座摩天大楼，建于 1930 年。

4. The 1940 Air Terminal Museum.
建于 1940 年的航空站博物馆。

5. Airport, Washington DC.
华盛顿机场。

门头门脸

FACADE & FRONT

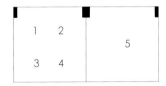

1/2. Kogen-Miller Studios.
高原·米勒工作室。

3. Financial entrance.
财政中心入口。

4. Desmond's Los Angeles-Entrance.
德斯蒙德洛杉矶入口。

5. Saint Joseph's Catholic Church Yorkville, William Schickel, Architect, New York City.
纽约约克维尔圣约瑟夫天主教教会。建筑师：威廉·史可尔。

门头门脸

FACADE & FRONT

门头门脸

1. Colombian Deco.
哥伦比亚装饰。

2. Montreal Art Deco.
蒙特利尔装饰艺术。

3. Capitol Annex.
国会大厦。

4. Brooklyn Industrial High School for Girls.
布鲁克林工业女校。

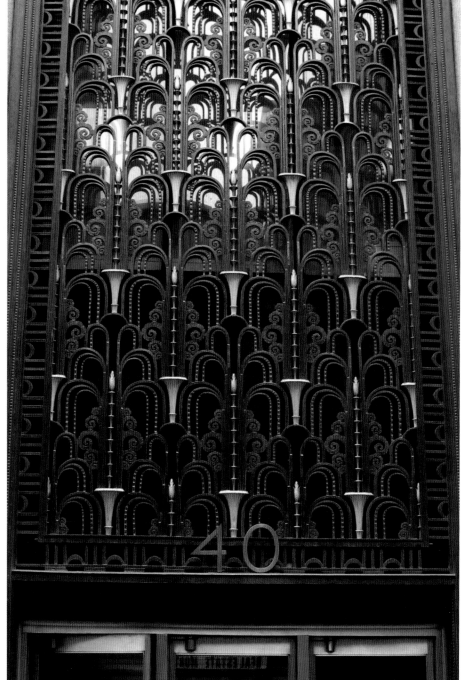

1. Third Man Records.
第三人纪录。

2. The Art Deco entrance of 40 East 34th Street.
东34号街40号的装饰艺术入口。

3. Directory.
名录。

4. Frist Auditorium.
弗里斯特礼堂。

5. Third Man Records.
三人唱片公司。

6. Grand Ocean Hotel, Saltdean. Ex Butlins holiday camp, now converted to luxury flats. The four main accommodation wings have been demolished, but the main block has been retained.
大洋酒店,前身是布特林度假令营,现在被改造成了奢华公寓。四个主要的住宿区已经拆除,但是主大厦依旧保留着。

门头门脸

FACADE & FRONT

门头门脸

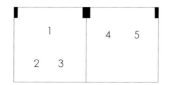

1. Art deco architecture at a residential area in Mexico City.
墨西哥城一个住宅区的装饰艺术建筑。

2. Designed by Emery Roth, originally intended to be a hotel.
由埃默里·罗斯设计,最初是一个酒店。

3. The Barber Institute of Fine Arts – inscription.
美术理发院的碑文。

4. National Breweries.
国家制酒厂。

5. Carnegie Library, Reims.
兰斯卡内基图书馆。

FACADE & FRONT

1	4	
2		
3	5	6

1. Sam Spade's Office is in the third floor.
萨姆·斯佩德的办公室在三楼。

2. Louisiana State Capitol.
路易斯安那州议会大厦。

3. The Film Center Building, at 630 Ninth Avenue between 44th and 45th Street in NY, is a 13-story office building catering to businesses involved in film, theatre, music and audio production and exploitation. It is listed on the National Register of Historic Places in 1984.
电影中心大厦，位于纽约第九大道 630 号与第四十四和第四十五街之间，是一座主营影视、戏剧、音乐和音频制作与开发业务的 13 层的办公大楼。该大厦在 1984 年被列入《国家史迹名录》。

4. State Theatre bronze doors.
国家剧院的铜制门。

5. The Daily Building, Napier.
纳皮尔日报大厦。

6. Federal Reserve Bank of Atlanta.
亚特兰大联邦储备银行。

门头门脸

FACADE & FRONT

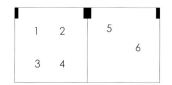

1/2. Art Deco facade.
装饰艺术风格立面。

3. 570 Lexington Ave night.
列克星敦大道 570 号夜景。

4. Entrance to Mr K's.
K 先生大厦入口。

5/6. 570 Lexington Ave art deco skyscraper, originally for RCA but eventually for GE.
列克星敦大道 570 号装饰艺术风格摩天大楼，原为美国广播唱片公司大楼，最终为通用电气公司所有。

门头门脸

FACADE & FRONT

门头门脸

1. Restored building, Napier.
纳皮尔翻新大楼。

2/3. Diamond Banqueting Suite - Skinner Street, Wolverhampton - former Odeon cinema.
伍尔弗汉普顿斯金纳街钻石宴会套房,原为国宾戏院。

4. The carvings over the door to the National Archaeological Museum in Tehran show a clear Art Deco influence, which is consistent with the museum's construction in the 1930s.
德黑兰国家考古博物馆大门的雕刻,显示了浓厚清晰的装饰艺术影响,与20世纪30年代博物馆的建造一致。

5. National Breweries.
国家制酒厂。

FACADE & FRONT

门头门脸

1/2/3/4. Art Deco facade.
装饰艺术风格立面。

FACADE & FRONT

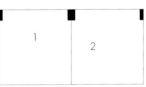

1. Art Deco facade.
装饰艺术风格立面。

2. Art Deco entrance to the Marine Building.
海事大厦装饰艺术入口。

门头门脸

门头门脸

1/2/3/4/5. Art Deco references, No.1 City Square, Park Row, Leeds.
利兹公园路城市广场 1 号装饰艺术风格。

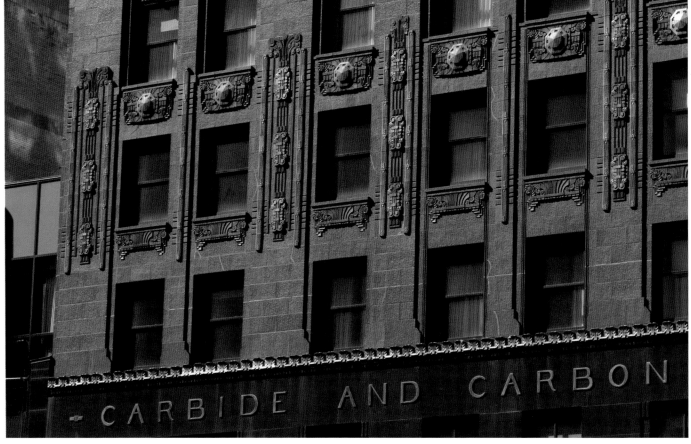

1/2. Carbide and Carbon Building Chicago.

芝加哥碳化物和碳大厦。

3. Completed in 1928, the Chicago Motor Club Building is an Art Deco landmark. The building is being restored and set to re-open in June 2015 as a Hampton Inn.

建于 1928 年的芝加哥汽车俱乐部大楼是装饰艺术风格建造的里程碑。大楼目前正在修建，计划于 2015 年 6 月作为一家汉普顿酒店重新开放。

门头门脸

FACADE & FRONT

1. Wonderful art deco building in Cincinnati.
辛辛那提美丽的装饰艺术风格大楼。

2. Royal Horticultural Hall.
英国皇家园艺大厅。

3. Rio de Janeiro Art Deco.
里约热内卢装饰艺术。

4. South Africa Art Deco.
南非装饰艺术。

5. Egyptian Theater De Kalb IL.
迪卡尔布埃及剧院。

门头门脸

FACADE & FRONT

		3	
1			
2		4	5

1/2. Prague Art Deco.
布拉格装饰艺术。

3. Egyptian Building Foyer Art Detail.
埃及剧院门厅艺术细部。

4/5. Deco Windows.
装饰窗户。

门头门脸

FACADE & FRONT

1. Eleven Madison Park.
十一麦迪逊公园。

2. Door back from terrace to interior marae, Te Papa.
蒂帕帕从阳台到室内毛利会堂的后门。

3. Napier New Zealand.
新西兰纳皮尔。

4. Durham Museum, Omaha NE.
奥马哈达勒姆博物馆。

5. Art Deco eagle.
装饰艺术风格老鹰。

门头门脸

064

FACADE & FRONT

1	2	3
		4 5

1. Encore.
安可大楼。

2. Empire State Building Front.
帝国大厦前门。

3. Restored building – no longer used as Public Trust Office.
修建的大楼,再也不是公共信托局。

4/5. Lapeyre Miltenberger Home for Convalescents.
拉佩尔·米尔滕伯格康复中心。

门头门脸

FACADE & FRONT

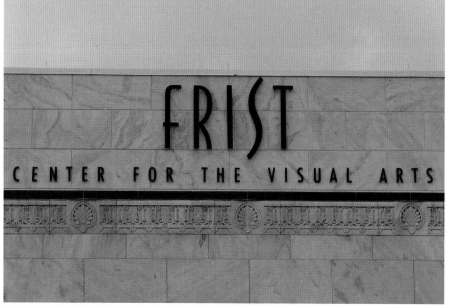

门头门脸

1. Frist Eagles.
弗里斯特大楼的老鹰。

2. Frist Center for the Visual Arts.
弗里斯特视觉艺术中心。

3. Frist Entrance.
弗里斯特视觉艺术中心入口。

4. Frist Center.
弗里斯特中心。

5. Paris Art Deco.
巴黎装饰艺术。

6. Downtown Miami Art Deco and 1920's buildings.
迈阿密市区装饰艺术风格和20世纪20年代建筑。

7. Facade of First American National Bank building in Port Townsend, Washington.
华盛顿汤森港的美国第一国民银行大楼立面。

FACADE & FRONT

1/2/3/4. General Laundry.
大众洗衣房。

门头门脸

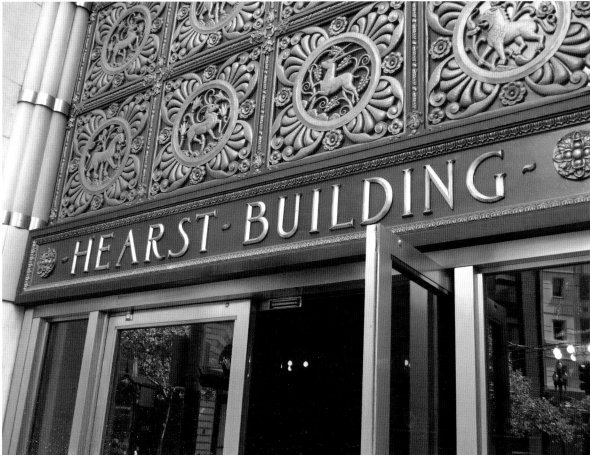
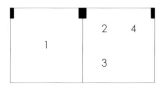

1/2/3/4. Hearst Building, San Francisco, California. The Financial District is a neighborhood in San Francisco, California, that serves as its main central business district.

加利福尼亚州洛杉矶赫斯特大厦。金融区就在附近，并且是加利福尼亚州洛杉矶的中心商务区。

门头门脸

1. Lakefront Airport.
湖畔机场。

2. Dallas Power & Light.
达拉斯电力与照明公司。

3. Lone Star Gas Co.
龙星天然气有限公司。

门头门脸

FACADE & FRONT

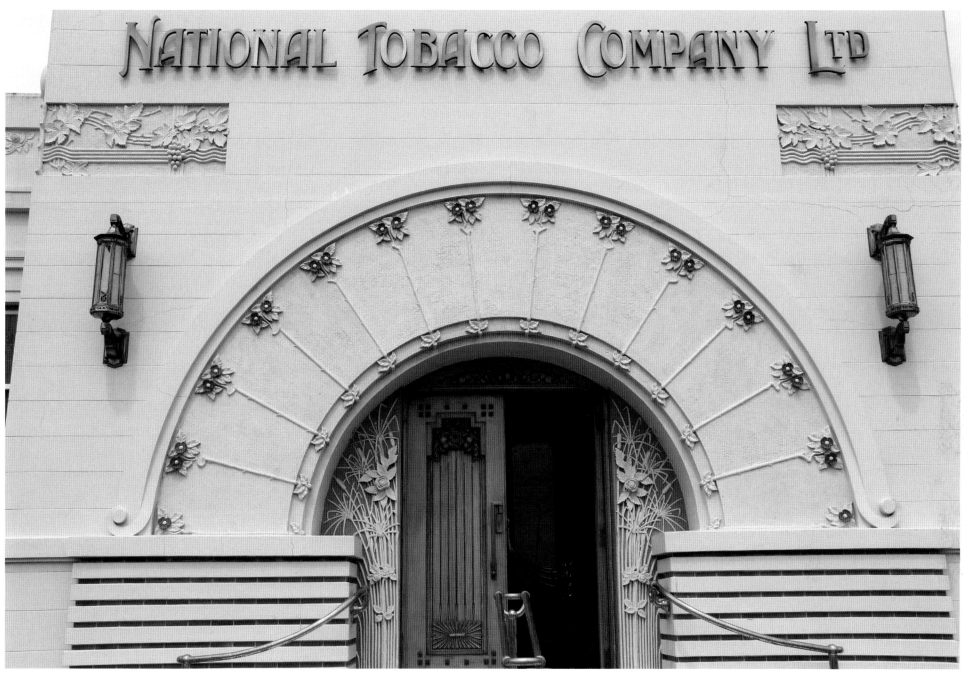

门头门脸

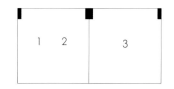

1. 183 East 34th Street (detail).
东 34 号街 183 号（细部）。

2. Chrysler Building.
克莱斯勒大厦。

3. Mixed entrance.
混合的入口。

FACADE & FRONT

1. Grand Ocean Hotel, Saltdean. Ex Butlins holiday camp, now converted to luxury flats. The four main accommodation wings have been demolished, but the main block has been retained.

大洋酒店，前身是布特林度假夏令营，现在被改造成了奢华公寓。四个主要的住宿区已经拆除，但是主大厦依旧保留着。

2. Malco.

马尔科。

3. One of the sights of the A40 going in to London, thanks to Tesco for this masterpiece of Art Deco extravagance.

A40 的其中一个景点将出现在伦敦，得益于特斯科公司为这个装饰艺术名作的付出。

4. Met Life North Entrance.

大都会人寿保险北门。

5. Art Deco Style Building Park Central Entrance in Miami Beach, Miami, Florida, USA.

美国弗罗里达州迈阿密港的中央公园酒店，是一座装饰艺术风格建筑。

1. Old building.
古老的大楼。

2. Museo de Antioquia.
安蒂奥基亚博物馆。

3. Splendid entrance to the Koppers Building in Pittsburgh, PA.
宾夕法尼亚州匹兹堡考伯斯大厦的辉煌入口。

门头门脸

门头门脸

1. Public Service Building.
公共服务大楼。

2. Hotel Gronstadt.
喀琅施塔得酒店。

3. Rockefeller Centre Mural.
洛克菲勒中心壁画。

4. Doorway into restored building.
修建大厦的入门。

1/2. Philadelphia.
费城。

3. San Remo Towers, Boscombe, Bournemouth.
164 luxury flats built in 1935-8 by American Hector O'Hamilton, in a Los Angeles Spanish flamboyant style.
伯恩茅斯博斯库姆圣雷莫塔。由美国人赫克托·欧汉密尔顿建于1935年8月的164个奢华公寓。风格为洛杉矶西班牙艳丽风格。

4. Pennsylvania Railroad Station.
宾夕法尼亚铁路站。

门头门脸

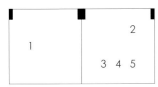

1/2. The Criminal Courts Building for the Parish of Orleans.
奥尔良教区的刑事法庭大楼。

3. 1 Wall Street.
华尔街一号。

4. Entrance to Art Deco Building, Albany, NY.
纽约州奥尔巴尼装饰艺术风格大厦的入口。

5. St. Peter's Church (Chicago Loop).
芝加哥环路圣彼得教堂。

门头门脸

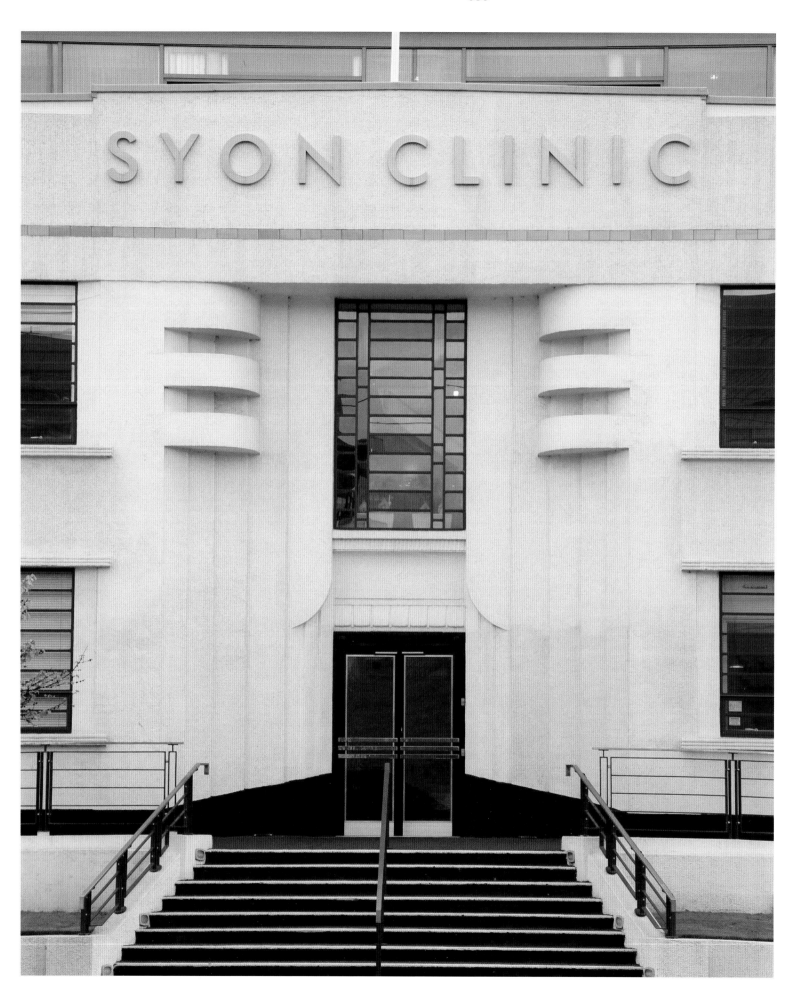

FACADE & FRONT

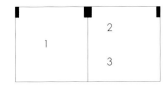

门头门脸

1. The Coty Building, Great West Road Brentford. Cosmetics, soap, lipsticks etc were made here until 1979. Taken over and beautifully maintained as the Syon Clinic.

布伦特福德大西路的科蒂大厦。在1979年前,化妆品、肥皂和口红曾经都在这里生产。后来被接管,并且作为西昂诊所美丽地保存了下来。

2. The Pyrene (fire extinguisher) Building, built in 1929 by Wallis, Gilbert and Partners (again!). Latterly used by Carillion (Westlink House), now stands empty.

芘(灭火器)大厦,由瓦利斯,吉尔伯特和合作人建于1929年。后来被卡利莲(西联大楼)所用,如今被空置着。

3. The old Daimler Hire Car Garage, now used by Mccann Erickson, advertising agency. An Art Deco classic!

旧的姆勒出租汽车车库现在被麦肯广告公司所用。这座大楼是装饰艺术的经典之作!

1/2/3. Tulane University.
杜兰大学。

4. The Milwaukee Art Museum.
密尔沃基艺术博物馆。

门头门脸

FACADE & FRONT

1. This attractive Art Deco building finished in Portland Limestone is located on an important corner site in Belfast. This Bank of Ireland branch closes off the long Royal Avenue vista.

这座用波特兰石灰岩粉饰的引人注目的装饰艺术风格大楼位于贝尔法斯特一个重要的街角底盘。这个爱尔兰银行分行封闭了长皇家大道远景。

2/3. Wrought Iron Door, Casablanca.
卡萨布兰卡锻铁大门。

4. Somerset House, Temple Street, Birmingham - relief of a sailing ship.
伯明翰庙街萨默塞特宫的帆船浮雕。

5. The Royal Shakespeare Theatre - Stratford-upon-Avon.
埃文河畔斯特拉特福皇家莎士比亚剧院。

6. Telephone House - Newhall Street, Birmingham.jpg
伯明翰纽荷尔街电话房。

门头门脸

1. Art Deco entrance.
装饰艺术风格入口。

2. Art Deco entrance.
装饰艺术风格入口。

3. Rio de Janeiro Art Deco.
日约热内卢装饰艺术。

4. Art Deco Door-This brass door accesses the left side of the lobby, which was later transformed into an employee cafeteria serving area.
装饰艺术大门——这个黄铜大门靠近左侧的大堂，大堂后期被改造成员工自助餐厅。

5. Art Deco entrance.
装饰艺术风格入口。

门头门脸

1. Wrought Iron Door, Casablanca.
卡萨布兰卡锻铁大门。

2. Wrought Iron Door.
锻铁大门。

3. Boulogne-Billancourt.
布洛涅 — 比扬古。

4. An art deco door iron decorated and framed by white marble.
铁艺装饰的装饰艺术大门,外框由白色大理石制成。

5. Wrought Iron Door, Casablanca.
卡萨布兰卡锻铁大门。

门头门脸

FACADE & FRONT

1. Boulogne-Billancourt.
布洛涅 — 比扬古。

2/3. Wrought Iron Door, Casablanca.
卡萨布兰卡锻铁大门。

4. Vintage bronze Art Deco door.
古老的青铜色装饰艺术大门。

5. Art Deco Entrance, Montevideo.
蒙得维的亚装饰艺术大门。

6. Wrought Iron Door, Casablanca.
卡萨布兰卡锻铁大门。

7. Tulsa Treasures.
塔尔萨瑰宝大厦。

8. Wrought Iron Door, Casablanca.
卡萨布兰卡锻铁大门。

FACADE & FRONT

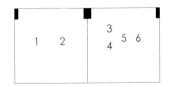

1. Bibliotheque Rosemont.
罗斯蒙特图书馆。

2. Havana.
哈瓦那。

3/4/5/6. These doors aren't IN Montreal but they were MADE there, designed by a famous Montrealer, and are a gift from Canada for a very famous building. Can you guess the building and the designer?

这些大门不是在蒙特利尔，但是是在蒙特利尔生产的，由一个出名的蒙特利尔人设计，是加拿大为一座非常出名的建筑所送的礼物。你能猜到这个建筑名和设计师名字吗？

门头门脸

FACADE & FRONT

1		5 6 7
2	3	8
	4	

1. Charity Hospital of Louisiana.
路易斯安那州的慈善医院。

2. Farmers and Mechanics Bank.
农民和工商银行。

3. Cadillac Hotel Miami Beach.
迈阿密海滩的凯迪拉克酒店。

4. Art Deco San Juan Hotel South Beach.
南海滩的装饰艺术风格圣胡安酒店。

5. Wrought Iron Door, Casablanca.
卡萨布兰卡锻铁大门。

6. Tokyo Metropolitan Teien Art Museum.
东京都庭园美术馆。

7. Toronto Art Deco.
多伦多装饰艺术。

8. Wrought Iron Door, Casablanca.
卡萨布兰卡锻铁大门。

门头门脸

FACADE & FRONT

门头门脸

1. 5410 Wilshire Boulevard.
威尔夏大道 5410 号。

2. New York Art Deco.
纽约装饰艺术。

3/4/5. American International Building.
美国国际大厦。

FAÇADE & FRONT

门头门脸

1. Elevator Sign DuPont Building Downtown Miami.
迈阿密市区的杜邦大厦的电梯标识。

2. Downtown Miami Art Deco and 1920's buildings.
迈阿密市区的装饰艺术和20世纪20年代的建筑。

3. Elevator Sign DuPont Building Downtown Miami.
迈阿密市区的杜邦大厦的电梯标识。

4/5. Havana Art Deco.
哈瓦那装饰艺术。

6. Entrance to an art Deco stylized building.
一座装饰艺术风格大楼的入口。

7. Art Deco elevator.
装饰艺术风格电梯。

1. New York Art Deco.
纽约装饰艺术。

2. Art Deco entrance.
装饰艺术入口。

3. New York Art Deco.
纽约装饰艺术。

4. Gulf Building.
海湾大厦。

5/6/7. Marine Building Vancouver.
温哥华海事大厦。

门头门脸

FACADE & FRONT

1. Art Deco doorway at the New Yorker Hotel.
纽约人酒店的装饰艺术门口。

2. Parliament of Poland.
波兰议会大厦。

3. Auditorium Deco.
礼堂装饰。

4/5/6/7. Newark Art Deco.
纽瓦克装饰艺术。

门头门脸

FACADE & FRONT

1. Rio de Janeiro Art Deco.
里约热内卢装饰艺术。

2. Entrance to the Frick Building in Pittsburgh (viewed from the lobby).
匹兹堡弗里克大厦的入口（从大堂望去）。

3. Carbide and Carbon Building Chicago 18.
芝加哥碳化物和碳大厦。

4. The original 1930s entrance hall to the Royal Shakespeare Theatre. The new build has kept some of the original Art Deco features.
建于20世纪30年代的皇家莎士比亚剧院入口大厅。新建筑保留了原建筑的一些装饰艺术特征。

5. A nice set of Art Deco doors at the Hoover Dam.
胡佛水坝的一套精美的装饰艺术大门。

门头门脸

FACADE & FRONT

1/2/3/4. Marine Building Vancouver.
温哥华海事大厦。

5. Art Deco designed doors and light fixtures in a Chicago building.
芝加哥一座大楼里的装饰艺术风格大门和照明。

门头门脸

1. Severance Hall.
塞弗伦斯音乐厅。

2. Wrought Iron Door, Casablanca.
卡萨布兰卡锻铁大门。

3. Bronze decorations of the doors of Duomo cathedral in Florence, Italy.
意大利佛罗伦萨多摩大教堂大门上的青铜装饰。

4. Havana Art Deco.
哈瓦那装饰艺术。

5. Lothian House Art Deco Door.
洛锡安大楼的装饰艺术大门。

6. Hostel Deco, Krakow.
克拉科夫装饰风格旅馆。

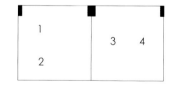

1. Rego Park Jewish Center.
雷哥公园犹太中心。

2. Art Deco entrance on an older building.
一座古老建筑的装饰艺术入口。

3/4. The Greater London House originally was built in 1926 as the Carreras Building (actually a Tobacco factory). The exterior has strong Art Deco influences. It was also said that it was inspired by an Egyptian temple to the cat-goddess Bubastis.

卡雷拉斯大厦（实际上是一个烟草厂）的大伦敦建于 1926 年。外部受装饰艺术的强烈影响。据说也受埃及猫头神巴斯特的宗教寺庙建筑的影响。

门头门脸

FACADE & FRONT

门头门脸

1. Havana.
哈瓦那。

2. Severance Hall.
塞弗伦斯音乐厅。

3. Art Deco entrance.
装饰艺术入口。

4/5. 1 Wall Street, Red Banking Room.
华尔街一号的红色银行房间。

FAÇADE & FRONT

1/2. Boulogne-Billancourt.
布洛涅 — 比扬古。

3. Paris Art Deco.
巴黎装饰艺术。

4. Cleveland Art Deco.
克利夫兰装饰艺术。

5. Tokyo Metropolitan Teien Art Museum.
东京都庭园美术馆。

6. Wrought Iron Door, Casablanca.
卡萨布兰卡锻铁大门。

7. New York Art Deco.
纽约装饰艺术。

门头门脸

1. Daily Telegraph Building, Napier.
纳皮尔每日电报大厦。

2. Havana Art Deco.
哈瓦那装饰艺术。

3. Childrens Hospital Broadway Plaza - Broadway - Casino.
博彩娱乐场百老汇广场儿童医院。

4/5/6. amazon.com sign, front entrance, windows.
亚马逊前门和窗户标识。

门头门脸

1/2/3/4. Cleveland Art Deco.
克利夫兰装饰艺术。

5/6/7. Cour d'Appel Montréal.
蒙特利尔上诉法院。

门头门脸

1. An art deco lobby in an office building.
一座办公楼的装饰艺术风格大堂。

2. The Black Madonna House.
黑色圣母之屋。

3. Musée national des Arts d'Afrique et d'Océanie.
非洲和大洋洲国家艺术博物馆。

4/5/6. Severance Hall.
塞弗伦斯音乐厅。

FACADE ORNAMENT
立面装饰

Art Deco（装饰主义）风格在拱券、柱式、雕塑等在立面设计上的运用源自古典主义的装饰题材和图案，但是，Art Deco（装饰主义）的雕塑和浅浮雕比起新古典主义更加样式化，趋于几何和简化，更具现代感，并且往往被赋予全新的题材。Art Deco（装饰主义）通过变形、简化、几何化处理，将古典装饰转变成了摩登装饰。几何图形屋顶对称简洁，立体感的浮雕形状是 Art Deco（装饰主义）的典型标志。

浓烈又精致的室内装饰品在艺术装饰品（如雕塑、壁画等）的选用上，雕塑通常选用富于曲线感的过渡作品，壁画通常选用抽象或富于异国情调的作品。

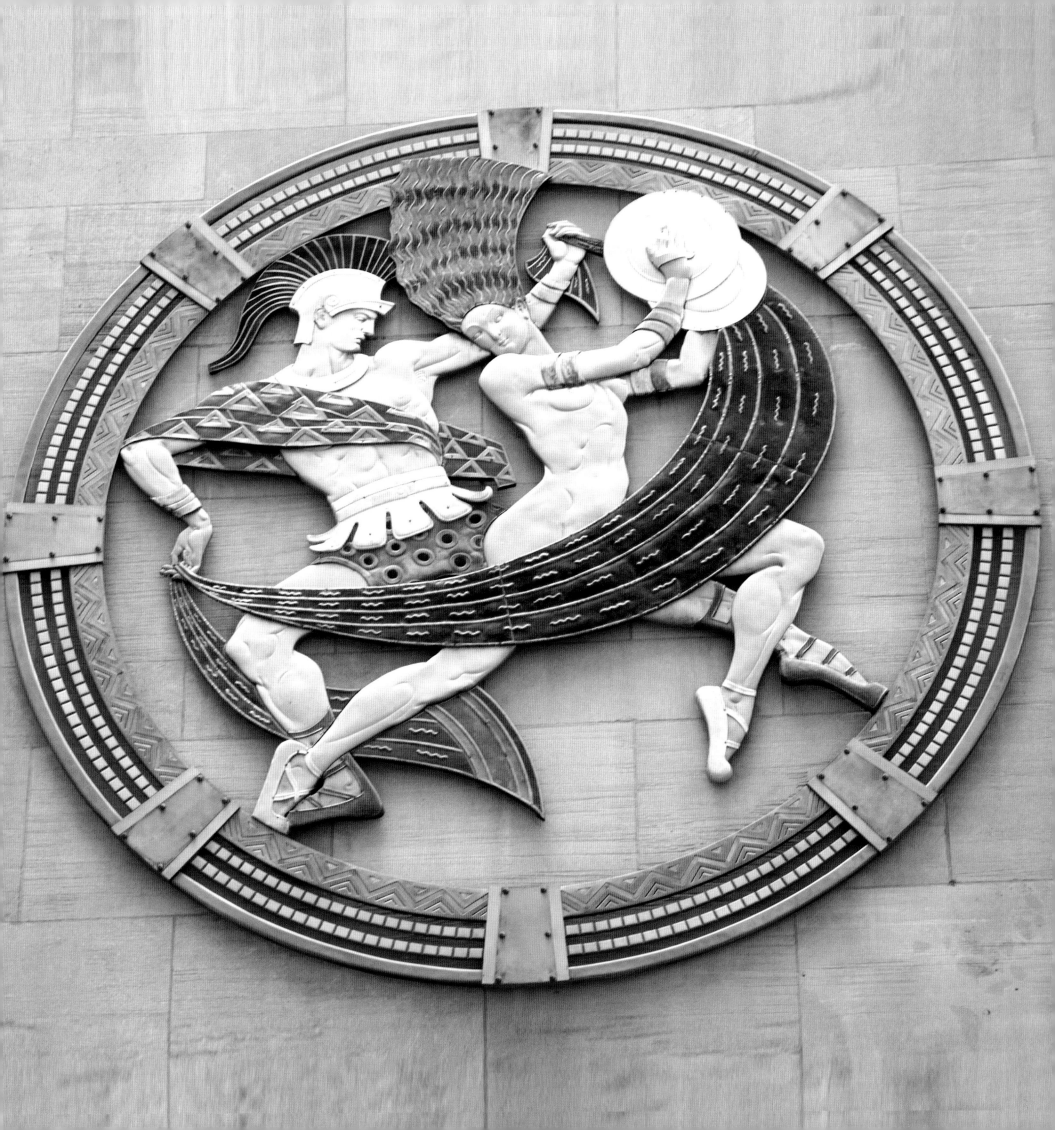

浮雕、雕塑

具有寓意式的装饰或花纹状的浮雕，是Art Deco（装饰主义）风格的特点之一，象征着新时代的黎明曙光。富有规则的几何纹路打造的浮雕效果运用于墙壁也是必要手法之一。Art Deco（装饰主义）风格极尽奢华，由高强度对比的黑、金演绎，给予视网膜的超强冲击。Art Deco（装饰主义）风格的图案纹路独具特色，锯齿图形和阶梯图形等，能够带来机械线条美感和装饰效果，震撼视觉神经。

浮雕、雕塑之人物篇

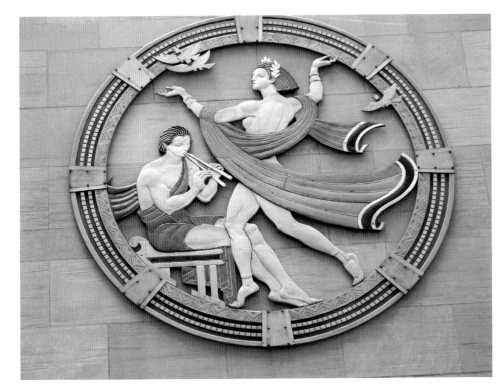

立面装饰

1. Art deco clock by Lee Lawrie located at the entrance of International Building at the Rockefeller center in New York.
李劳列设计的装饰艺术钟表，装饰在纽约洛克菲勒中心的国际大厦入口上。

2/3/4. New York Art Deco.
纽约装饰艺术。

5. The Municipal House in Prague.
布拉格市政大厦。

FAÇADE ORNAMENT

立面装饰

1/2/3/4. Snowdon Theatre, Dancer.
斯诺登剧院的舞者。

5/6/7. Main foyer in the Express Building, Fleet St.
舰队街快递大厦的主门厅。

FACADE ORNAMENT

立面装饰

1. The Fifth Avenue.
第五大道。

2. Rockefeller Center, Manhattan, New York.
纽约曼哈顿洛克菲勒中心。

3/4. New York Art Deco.
纽约装饰艺术。

FACADE ORNAMENT

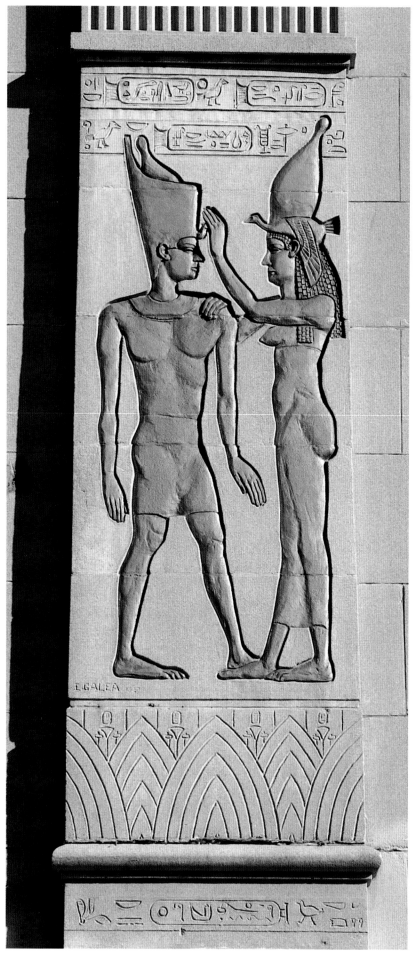

1. Empress Theatre.
皇后剧院。

2. Havana.
哈瓦那。

3. Hydro Power, bank of Canada.
加拿大银行电力大厦。

4. Tokyo Metropolitan Teien Art Museum.
东京都庭园美术馆。

5. Bringing up the guns, Mont St. Quentin.
蒙特圣昆廷造枪雕塑。

立面装饰

FACADE ORNAMENT

1. "Ball Playing Ceremony" relief, circa 380-246 B.C.
约公元前 246 至 380 的控球仪式浮雕。

2/3. Bank of Canada.
加拿大银行。

4. Cabot Memorial.
卡博特纪念馆。

5/6/7/8. Supreme Court.
最高法院。

立面装饰

FACADE ORNAMENT

1. Washington DC internal capitol dome painting.
华盛顿内部国会大厦圆顶画。

2. Memories not to lose.
不曾遗忘的记忆。

3. Saint Jean d'Angely - art deco.
圣让安格伊装饰艺术雕塑。

4. Capitol Annex.
国会大厦雕塑。

5. Unilever House.
联合利华大厦。

立面装饰

142

FACADE ORNAMENT

立面装饰

1/2. Vatican Inside Ornate Ceiling with Scullpture of Christ with Disciples.
梵蒂冈里面华丽的天花板与耶稣与其弟子的雕塑。

3. Westminster Abbey North Entrance.
威斯敏斯特教堂北门。

4/5/6. Tulane University.
杜兰大学。

FACADE ORNAMENT

立面装饰

1. Saint Ignatius Loyola Catholic Church Bronze Door by Long Island Bronze Company, New York City.
纽约长岛青铜公司制造的圣依纳爵罗耀拉天主教会的青铜大门。

2. Albert Memorial Bridge.
艾伯特纪念桥。

3. Paramount Theater.
派拉蒙剧院。

4. Wall decoration.
墙壁装饰。

5. Saint Thomas Episcopal Church World War I Memorial by Lee Lawrie, New York City.
由李劳列设计的纽约圣托马斯圣公会第一次世界大战纪念碑。

6. Paramount Theater.
派拉蒙剧院。

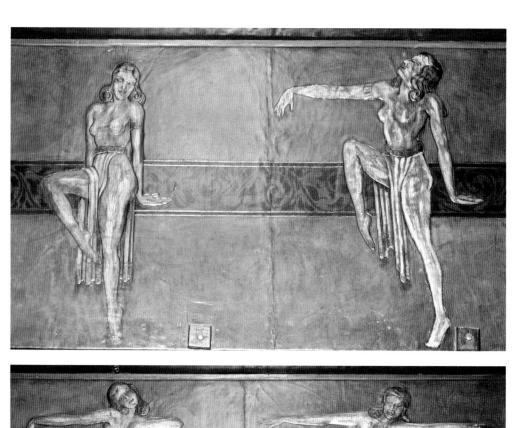

1/2/3. Teatro America, Havana.
哈瓦那美国剧院。

4. Queen Mary Hotel.
玛丽皇后酒店。

5. Greyhound terminal.
洛杉矶"灰狗"汽车总站。

6. Saint-Quentin Art Deco.
圣康坦装饰艺术。

7. Art deco interior of the Alameda Theatre.
阿拉米达剧院的装饰艺术风格室内。

立面装饰

1/2. The Noguchi Museum.
野口勇博物馆。

3. Relief in Carew Tower Arcade.
卡鲁塔商场浮雕。

4/5/6/7. New York Art Deco.
纽约装饰艺术。

立面装饰

浮雕、雕塑之图案篇

1/2. National Breweries.
国家酿酒厂。

3. Café Champs-Elysées.
香榭丽舍大街咖啡厅。

4. New York Art Deco.
纽约装饰艺术。

5. Ministry of Education, Warsaw.
华沙教育部。

6. Detail of Pewter Art Deco Tile from Historic Washington, DC Building.
历史悠久的华盛顿大楼的锡制的装饰艺术瓷砖细部。

立面装饰

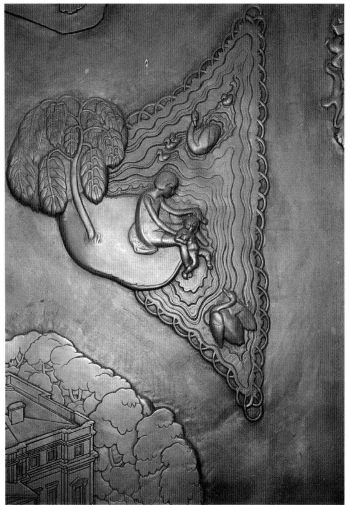

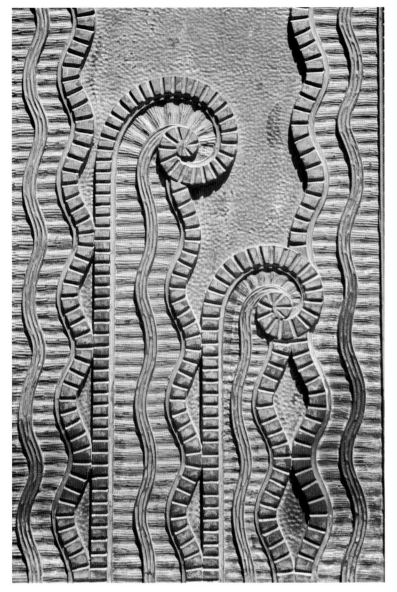

1/2/3. London Art Deco.
伦敦装饰艺术。

4. Art Deco swirling plant pattern stone relief architectural decoration on building exterior.
建筑外面的旋转的植物图案石雕建筑装饰。

5. New York Art Deco.
纽约装饰艺术。

6. Art Deco fish.
装饰艺术鱼图案。

立面装饰

FACADE ORNAMENT

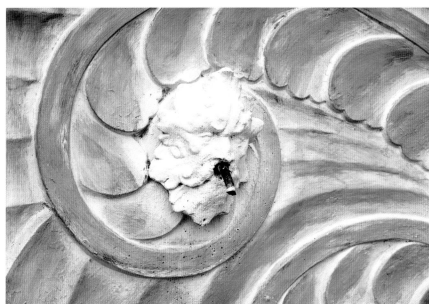

1. Florida Art Deco.
弗罗里达装饰艺术。

2. Art Deco wall.
装饰艺术墙壁。

3. Former Institute of Anatomy.
前身为解剖学研究所。

4. One Wall Street, originally the Irving Trust Company Building.
华尔街一号，前身为欧文信托公司大厦。

5. Zeppelin relief on the Sinclair Centre.
辛克莱中心的齐柏林浮雕。

6. Vancouver Art Deco.
温哥华装饰艺术。

7. Toronto Art Deco.
多伦多装饰艺术。

8. A colorful art deco design on a building in the South Beach area of Miami Beach showcases bold, bright colors.
迈阿密滩的南部海滩的一座色彩鲜艳、大胆的装饰艺术设计。

9. An architectural detail from a fountain in front of a South Beach hotel shows Poseidon against an aqua blue background.
南部海滩一座酒店前的喷泉建筑细部，展示了水蓝色背景下的波塞冬。

立面装饰

FACADE ORNAMENT

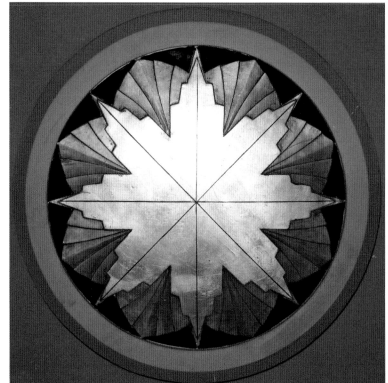

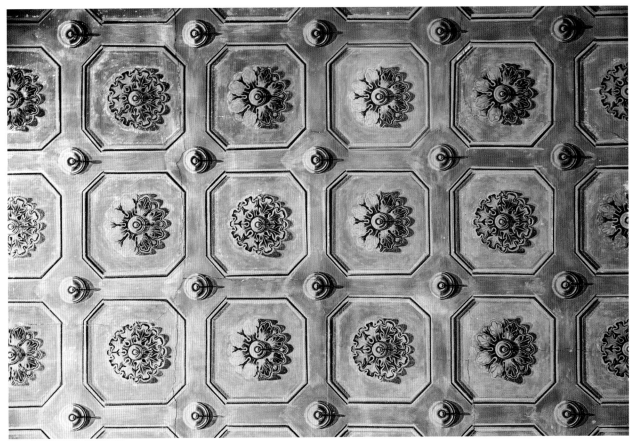

1. Texas & Pacific Passenger Terminal.
德州及太平洋客运码头。

2. Art Deco Montreal.
蒙特利尔装饰艺术。

3. Snowdon Theatre, Art Deco Flower.
斯诺登剧院装饰艺术花纹。

4. Rose art-deco detail from old book.
古门上的装饰艺术玫瑰图案细部。

5. Bronze art deco wall taken out of a lift.
电梯里的青铜装饰艺术墙。

6. Interior of historical building – hand-painted ceiling.
历史悠久的大楼室内的手绘天花顶。

立面装饰

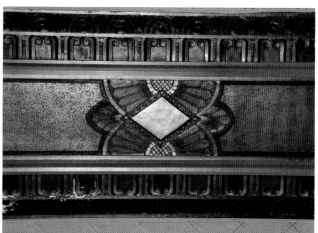

1/2/3/4/5/6/7/8/9. Chateau Theatre.
城堡剧院。

立面装饰

FACADE ORNAMENT

1/2. Hoover Dam.
胡佛水坝。

3/4. Paris Galeries Lafayette.
巴黎拉斐特百货公司。

立面装饰

FACADE ORNAMENT

立面装饰

1/2/3. Miami Art Deco.
迈阿密装饰艺术。

4/5/6. Fort Anne.
安妮堡。

7. Paramount Theater.
派拉蒙剧院。

FACADE ORNAMENT

1/4. New York Art Deco.
纽约装饰艺术。

2. Nelson Mandela Plenary Hall.
纳尔逊·曼德拉全会厅。

3. Severance Hall.
塞弗伦斯音乐厅。

5. Grand Ocean Hotel, Saltdean.
Ex Butlins holiday camp, now converted to luxury flats.
大洋酒店，前身是布特林度假夏令营，现在被改造成了奢华公寓。

立面装饰

立面装饰

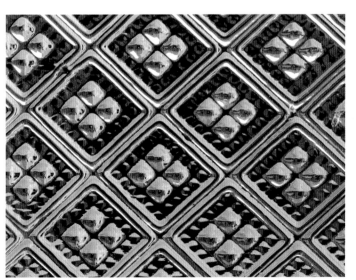

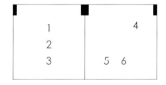

1/2/3/4. Art Deco animals.
装饰艺术动物图案。

5. Beautiful vintage beveled surface spot-illuminated wallpaper with metallic shine.
美丽的老式斜面上闪烁金属光芒的壁纸。

6. Ceramic old style on the wall.
墙上的老式陶瓷。

FACADE ORNAMENT

1	2		6	7
3	4	5		8

1/2. Daily Express Art Deco.
每日快报大厦内装饰艺术。

3. Art Nouvau detail Bat wing and flowers.
新艺术风格蝙蝠翅膀和花朵图案。

4. Art Deco Staircase, Chicago Daily News Bldg, 1929.
建于1929年的芝加哥每日经济新闻大厦的装饰艺术楼梯。

5. Port Wakefield. Pareora homestead. Art Nouveau wall panels done by Gustav Gebhardt after 1897.
韦克菲尔德港口，安全之家。1897年后古斯塔夫·格布哈特制作的新艺术风格墙板。

6. Art Deco Owl.
装饰艺术猫头鹰。

7. Art Deco letters.
装饰艺术风格书信。

8. Egyptian Building Baruch Auditorium Wall Sconces.
埃及建筑巴鲁克礼堂的壁灯。

立面装饰

FACADE ORNAMENT

立面装饰

1		5	6
2	4		
3			7

1. Philadelphia Custom House details.
费城海关大楼细部。

2. Art Deco building detail from California.
加利福尼亚装饰艺术风格建筑细部。

3. Port Townsend Art Deco.
汤森港装饰艺术。

4. Details of Villa Nobel, house of the famous chemist Alfred Nobel.
诺贝尔的别墅细部图,著名化学家阿尔弗雷德·诺贝尔的家。

5. Elevator Sign DuPont Building Downtown Miami.
迈阿密市区杜邦大厦的电梯标识。

6. Golden Iron work.
金色铁艺。

7. A polished gold neoclassical plaque on a building in downtown Chicago.
芝加哥市中心建筑上的抛光镀金新古典牌匾。

172

FACADE ORNAMENT

立面装饰

1. The Geary Theater.
基尔剧院。

2. Cleveland Art Deco.
克利夫兰装饰艺术。

3. Building decoration.
建筑装饰。

4. The Frist Center for the Visual Arts. The building was a New Deal project and opened in 1934 as the main Nashville US Post Office.
弗里斯特视觉艺术中心。该建筑是一个新政项目,并在 1934 年开业,作为纳什维尔主要的美国邮政局。

1. Chanin Building's striking relief panelling catches the cool autumn sun along Lexington Avenue.
查宁大厦引人注目的浮雕镶板反射了列克星敦大道秋天的太阳光。

2. Frise de liserons.
弗里斯兰旋花。

3. At the National Archaeological Museum, Tehran.
德黑兰国家考古博物馆。

4. Art deco decoration.
装饰艺术装饰品。

5. Columbus Circle, 59th Street Station.
第59街站哥伦布圆环。

立面装饰

壁画

Art Deco（装饰主义）的重点不在于华丽而在于艺术美感的呈现。1920～30年代风靡纽约的 Art Deco（装饰主义）风格，不仅金碧辉煌——代表那个资本主义乐观向上的时期，而且艺术含金量极高：Art Deco（装饰主义）地标之一洛克菲勒中心，里里外外充满彩色的壁画是艺术的杰作；另一个地标是 Waldorf Astoria 酒店，不仅壁画地板画名作连绵，Art Deco（装饰主义）风格浮雕，甚至完美覆盖从邮政信箱、电梯门和客房门把手等等一切细节。

1. Saint-Quentin Post Office.
圣康坦邮局。

2/3. Cleveland Art Deco.
克利夫兰装饰艺术。

4/5/6. Manchester Unity Building.
曼彻斯特联合大厦。

立面装饰

立面装饰

1/2/3/4. Rome Art Deco.
罗马装饰艺术。

5/6. Havana Art Deco.
哈瓦那装饰艺术。

180

FACADE ORNAMENT

立面装饰

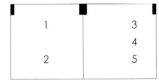

1/2/3/4/5. Saint-Quentin Post Office.
圣康坦邮局。

FACADE ORNAMENT

1. Saint Ignatius Loyola Catholic Church Apse, New York City.
纽约圣依纳爵罗耀拉天主教教会后殿。

2. Art Deco Montevideo.
蒙得维的亚装饰艺术。

3. Newark Art Deco.
纽瓦克装饰艺术。

4. Park Güell, Barcelona.
巴塞罗那桂尔公园。

立面装饰

FACADE ORNAMENT

立面装饰

1/2/3. "I'll Put a Girdle Round About the Earth".
"我要给地球绑上一个束带"。

4/5. New York Art Deco.
纽约装饰艺术。

6. Louisiana State Capitol.
路易斯安那州议会大厦。

立面装饰

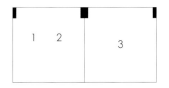

1. Danas Brussels courthouse.
达纳斯布鲁塞尔法院。

2. Dedicatoria a los arquitectos de la Gran Plaza Bruxelles, Belgique.
比利时布鲁塞尔为大广场的建筑师的敬辞。

3. Adorn in Novokuznetskaya metro station.
车站装饰。

1/2. Facade decoration.
立面装饰。

3. Inscriptionon the wall.
墙上碑文。

4. El Museo de Antioquia.
安蒂奥基亚省博物馆。

立面装饰

1. In 1932, Napier was leveled by an earthquake. It was rebuilt in Art Deco style, and most of those buildings stand today.
1932年，纳皮尔被地震夷为平地。城市以装饰艺术风格重建，而今大部分的建筑依然保留着。

2. Vitrolite Wall Colony Hotel.
殖民地酒店瓷板墙。

3. Rio de Janeiro Art Deco.
里约热内卢装饰艺术。

4. Montreal Art Deco.
蒙特利尔装饰艺术。

5/6. Sportsman Pub.
运动员酒吧。

立面装饰

FACADE ORNAMENT

立面装饰

1/2/3/4/5. Louisiana State Capitol.
路易斯安那州议会大厦。

铁艺

传统的铁艺主要运用于建筑、家居、园林的装饰，在法国、英国、意大利、瑞士、奥地利等欧洲的国家里装饰的运用都颇为广泛，从皇家到民居，从园林到庭院，从室内楼梯至室外护栏，形态各异，精美绝伦的装饰比比皆是。

从它们的线条、形态和色彩几方面比较，具有独特风格和代表性的是英国和法国的铁艺，而两国铁艺又各成风格，英国的铁艺整体形象庄严、肃穆、线条与构图较为简单明朗，而法国的铁艺却充满了浪漫温馨、雍容华贵的气息。如果说英国的铁艺像一个英俊倜傥的绅士，那么法国的铁艺则似一位华冠锦带的皇子。

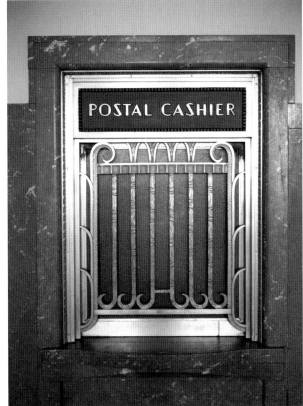

立面装饰

1. Rose iron works.
玫瑰花纹铁艺作品。

2. Southern California Edison Building.
南加州爱迪生公司大楼。

3. Sort of an Art Deco Christmas Card-A decorated door at street level on the Cincinnati Bell building.
街道上辛辛那提贝尔建筑的一个装饰门上的一种装饰艺术风格的圣诞卡。

4. This Heroes' Medal, which was the highest honour of Communist Ethiopia, weighs 700 kilograms and has a diameter of 2.7 m.
这个英雄的勋章是埃塞俄比亚共产党最高的荣誉，重700千克，直径2.7米。

5. Postal cashier.
邮政窗口。

6. Palace of Fine Arts.
美术宫。

7. Art Deco decoration.
装饰艺术装饰品。

196

FACADE ORNAMENT

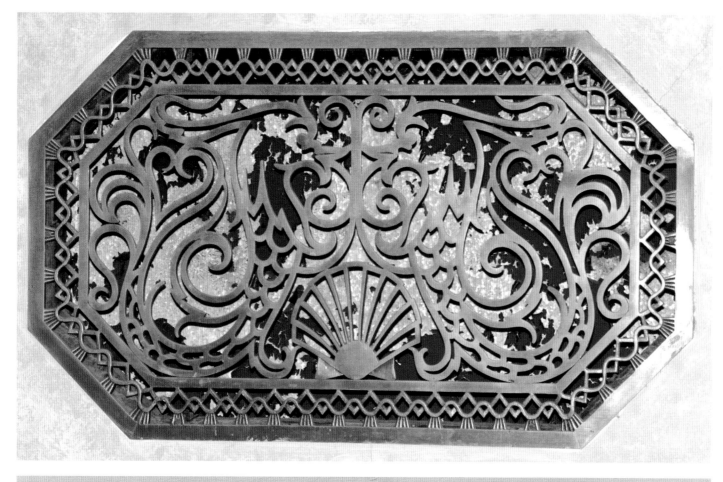

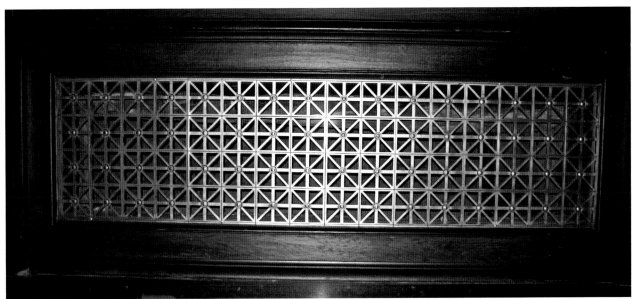

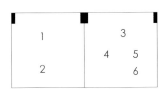

1. Chateau Laurier swimming pool.
劳雷尔城堡游泳池。

2. Moscow Metro.
莫斯科地铁。

3/4. Art Deco circles.
装饰艺术圆形图案。

5/6. Cleveland Art Deco.
克利夫兰装饰艺术。

立面装饰

FACADE ORNAMENT

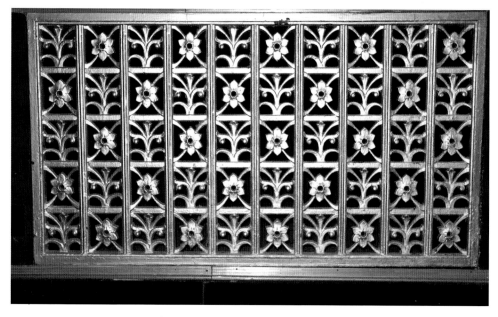

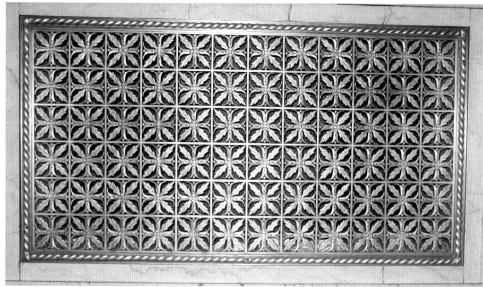

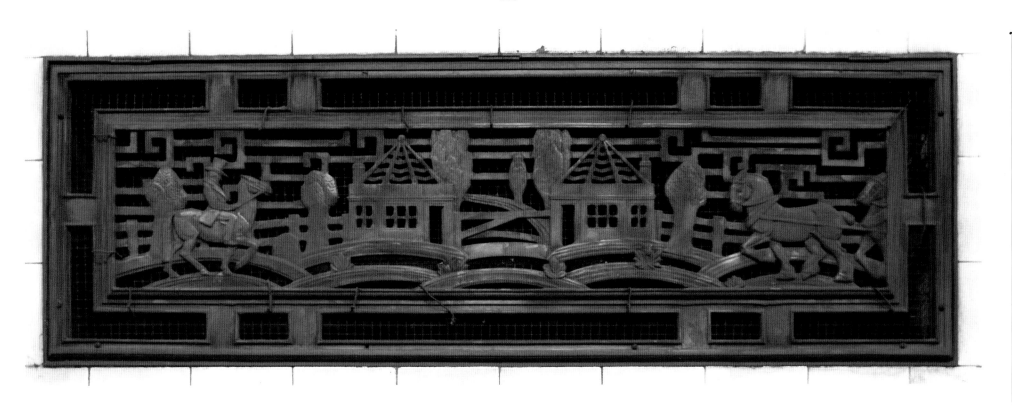

|1|4|6|
|2|5| |
|3| |7|

1/2/3/4/7. Cleveland Art Deco.
克利夫兰装饰艺术。

5. Art Deco detail.
装饰艺术细部。

6. Art Deco ventilation grille, depicting the local area theme.
艺术装饰通风格栅，体现了局部区域的主题。

立面装饰

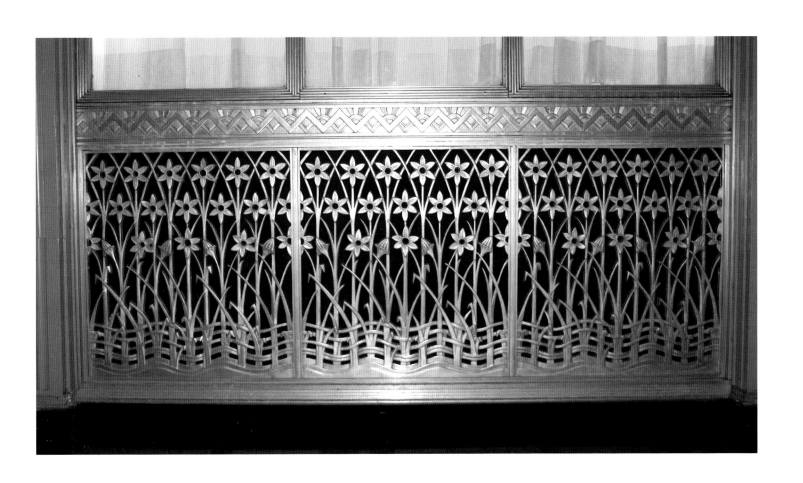

FACADE ORNAMENT

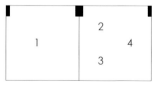

1. Capitol Annex.
国会大厦装饰。

2. Cleveland Art Deco.
克利夫兰装饰艺术。

3. Lasalle – lobby.
拉萨尔大厦大堂。

4. New York Art Deco.
纽约装饰艺术。

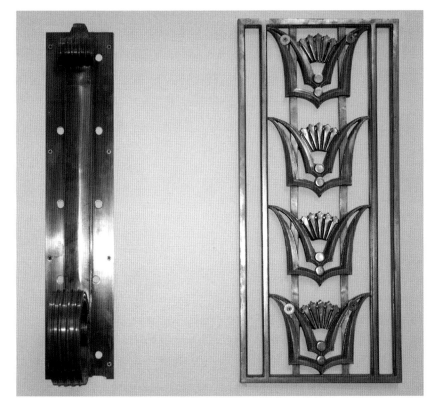
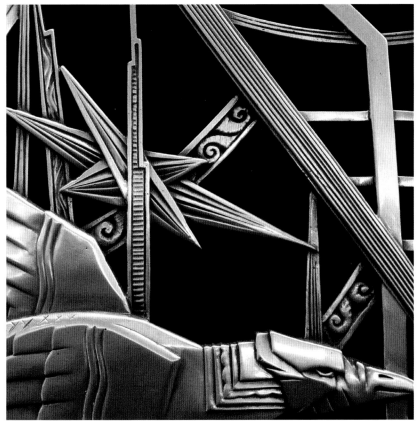

立面装饰

FACADE ORNAMENT

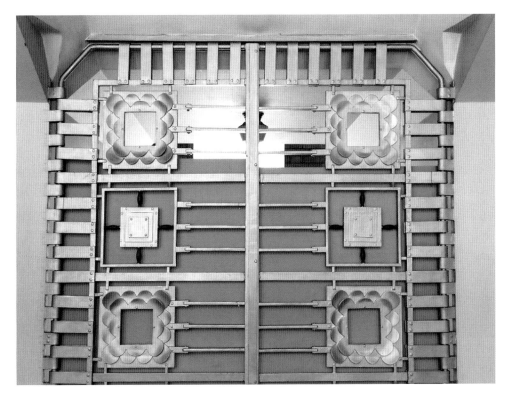
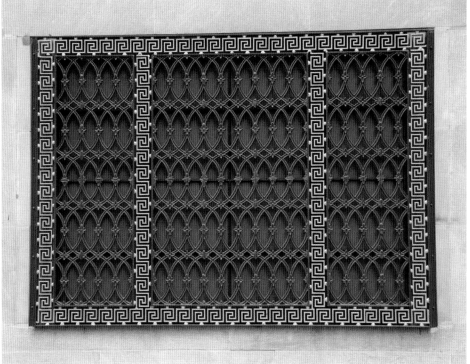

立面装饰

1. New York Art Deco.
纽约装饰艺术。

2/3. Durham Museum, Omaha NE.
奥马哈达勒姆博物馆。

4. Deco metal works.
装饰铁艺作品。

5. Seven Park Avenue Manhattan (detail).
曼哈顿七星公园大道（细部）。

6. Deco window.
装饰窗户。

CEILING
天花板

辐射状的太阳花纹理天花板是特别具有代表性的 Art Deco（装饰主义）机械美学，像是黎明的曙光在屋顶绽放，夸张而极富存在感。华丽的彩色玻璃打造的天花板，细节处的质感再次将整个空间的风格高贵和典雅一一呈现。

天花上点缀的彩色弧形玻璃管的艺术吊灯，线条柔美，色彩艳丽。这里设置着物业服务总台，舒适的围合形沙发，厚实的地毯。蓝色调的沙发、地毯及墙面的巨幅挂画，与整个空间的暖金色调，形成强烈的色彩碰撞。

Art Deco（装饰主义）是永恒艺术的永恒魅力。

CEILING

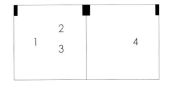

1. Interior dome of the Pennsylvania State Capitol building in Harrisburg, Pennsylvania.
宾夕法尼亚州哈里斯堡的宾夕法尼亚州议会大厦的内部顶灯。

2. Capitol dome from the rotunda of the Old Mississippi State Capitol building in Jackson, Mississippi.
在密西西比州杰克逊的旧密西西比州议会大厦的圆形大厅。

3. Interior dome from the rotunda floor of the Mississippi State Capitol building in Jackson, Mississippi.
在密西西比州杰克逊的旧密西西比州议会大厦的内部顶灯。

4. Guardian Building.
嘉德大厦。

天花板

CEILING

1. Interior, Library of Congress, Washington, DC, showing stained glass dome and fresco ceiling.
华盛顿特区的国会图书馆室内的彩色玻璃穹顶和壁画天花板。

2. Paramount Theater.
派拉蒙剧院。

3. Paris Art Deco.
巴黎装饰艺术。

4. Oklahoma state capitol building dome in Oklahoma City, Oklahoma.
俄克拉何马州国会大厦的穹顶。

5. Cincinnati Union Terminal.
辛辛那提联盟航站楼。

天花板

1/2. Silesian Parliament.
西里西亚议会。

3/4/5. Washington DC internal capitol dome view.
华盛顿国会大厦室内穹顶。

天花板

1. Interior of capitol dome in St. Paul, Minnesota.
美国明尼苏达州圣保罗国会大厦室内穹顶。

2. A colorful Italian Renaissance fresco on the arched ceiling of an ancient palace.
古老宫殿的拱形天花板印刻着意大利文艺复兴时期色彩明艳的壁画。

3. The golden ceiling in the Hermitage Museum, St. Petersburg, Russia.
埃尔米塔日博物馆的金色天花板。

4. Cleveland Art Deco.
克利夫兰装饰艺术。

5. Saint Ignatius Loyola Catholic Church Ceiling, New York City.
美国纽约圣依纳爵罗耀拉的天主教天花板。

天花板

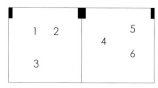

1/2. NBC Tower.
美国全国广播公司芝加哥办公塔楼。

3. Carbide and Carbon Building Chicago.
芝加哥联碳大楼。

4. Disney's Hollywood Studios.
迪士尼好莱坞影城。

5. Interior dome of the Michigan State Capitol building.
密西根州议会大厦的室内穹顶。

6. Dixie Terminal.
迪克西终点站大楼。

天花板

CEILING

1/2/3. Fisher Building designed by architects Albert Kahn.
底特律渔人大厦由德裔美国建筑师阿尔伯特·卡恩设计。

4/5/6. Interior of the Fisher Building.
渔人大厦的室内设计。

天花板

CEILING

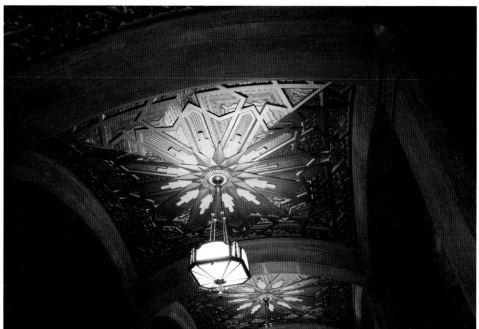

天花板

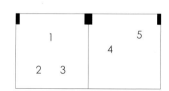

1. The ceiling of Fisher Building designed by architects Albert Kahn.
由建筑师阿尔伯特·卡恩设计的渔人大厦内天花板。

2. Miramare castle near Trieste, Italy.
意大利里雅斯特的地中海边的白色城堡。

3. JPMorgan Chase Tower (Houston).
美国休斯敦摩尔根大通大厦。

4/5. Galeries Lafayette Department Store in Paris.
法国巴黎老佛爷百货。

CEILING

天花板

1/2. Havana Art Deco.
哈瓦那装饰艺术。

3. Queen's Theatre London.
伦敦女王剧院。

4. Interior view of dome from ground floor in Arkansas State Capitol.
阿肯色州政府室内第一层楼穹顶。

5. Oviatt penthouse.
奥维亚特阁楼。

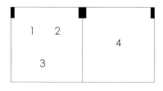

1/2/3. Snowdon Theatre.
斯诺登剧院。

4. Downtown Miami Art Deco and 1920's buildings.
20世纪20年代迈阿密市中心建筑的装饰艺术。

天花板

CEILING

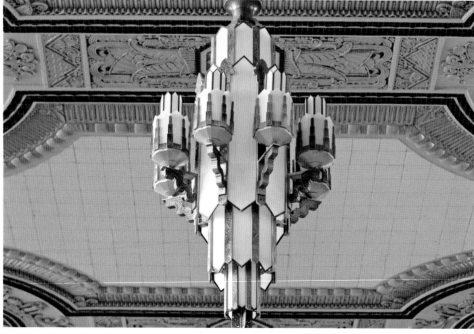

天花板

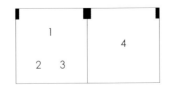

1/2/3/4. The Texas & Pacific Station building.
沃思堡的德克萨斯和太平洋铁路客运站大厦。

CEILING

天花板

1/2/3/4/5. Severance Hall.
塞弗伦斯音乐厅。

6/7. Ceilings inlaid with patterns.
镶嵌花纹的天花板。

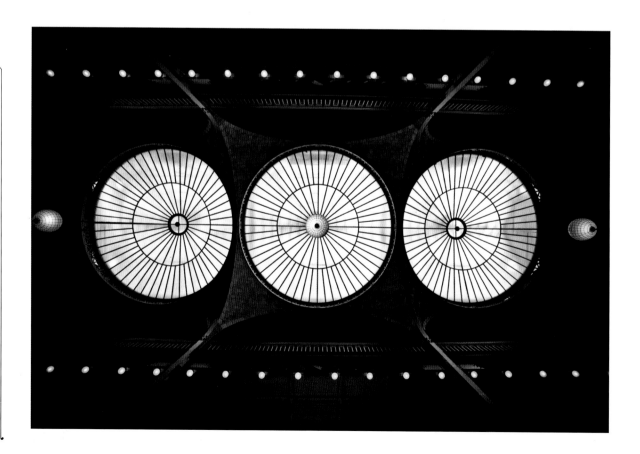
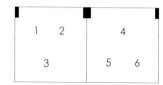

1/2. The Guggenheim Museums.
索罗门·古根汉美术馆。

3. Art Deco dome in a building in San Francisco.
三藩市一座大厦内的装饰风艺术穹顶。

4/5/6. Standard Building 1924.
美国标准大厦,建于 1924 年。

天花板

CEILING

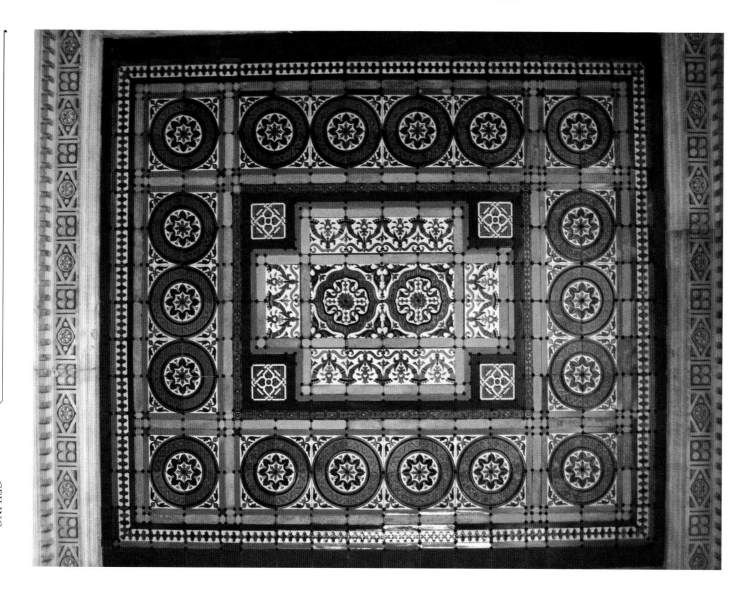

天花板

1. Foreign & Commonwealth Office.
外交及联邦事务部。

2/3. Paramount Theater.
西雅图派拉蒙剧院。

4. Art Deco dome in a building in Rome.
罗马一座大厦内的装饰艺术穹顶。

CEILING

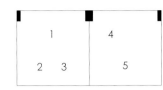

1. The Studebaker Theatre, also known as Studebaker Hall is housed in the The Studebaker Building in Chicago and was dedicated in 1898.
斯图特贝克剧院，也被称为史蒂倍克堂，坐落于芝加哥斯图特贝克大厦，于 1898 年对外开放。

2/3. Prague Art Deco.
布拉格装饰艺术。

4. Harrods Food Hall, Brompton Road, London.
伦敦布朗普顿路的一家哈罗德食品馆。

5. Art Deco dome in a building in Rome.
罗马一座大厦内的装饰艺术穹顶。

天花板

1. The Civic Opera house.
市政歌剧院。

2. The domed ceiling of temples in Wilmette Illinois USA.
位于美国伊利诺斯州的维尔梅特寺庙内的圆顶天花板。

3. Judge Sarah T. Hughes' courtroom.
大法官莎拉·T·休斯的审判室。

4/5. Lakefront Airport.
美国新奥尔良湖畔机场。

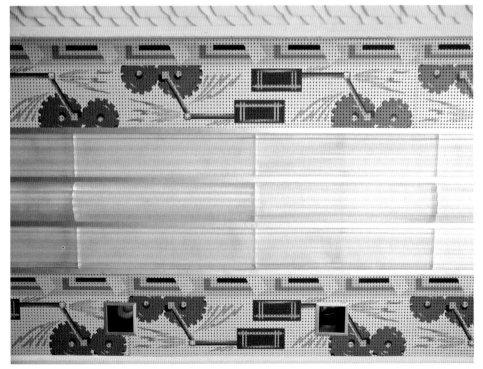

天花板

1. New York Art Deco.
纽约装饰艺术。

2/3/4/5. Daily Express Art Deco.
英国每日快报大厦内的装饰艺术。

天花板

CEILING

天花板

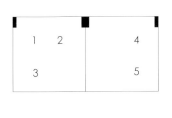

1. Newark, Art Deco.
纽瓦克装饰艺术。

2. Inside the Empire State Building.
纽约帝国大厦室内。

3. Prague coat of arms in Obecní dum.
布拉格市民会馆内的盾形纹章天花板。

4. Bellas Artes.
墨西哥城国家美术馆。

5. The Eye of the Dome.
穹顶之眼。

1. Tulsa Treasures.
塔尔萨瑰宝大厦。

2. Havana Art Deco.
哈瓦那装饰艺术。

3/4. Saint Ignatius Loyola Catholic Church Baptistry Semi Dome by Louis Comfort Tiffany, New York City.
纽约罗耀拉圣依纳爵天主教教堂的穹顶，由美国艺术家路易斯·康福特·蒂芙尼设计。

天花板

1. Hall of Mirrors.
镜厅。

2. Ornate vintage gold ceiling with a hanging lamp.
奢华而复古的金色吊灯天花板。

3. The Petronas Towers, also known as the Petronas Twin Towers, are twin skyscrapers in Kuala Lumpur, Malaysia.
吉隆坡石油塔，又名吉隆坡石油塔，是马来西亚吉隆坡的摩天大楼。

4. Gothic foyer ceiling.
哥特式门厅的天花板。

5. Cafe du Palais.
咖啡厅杜宫。

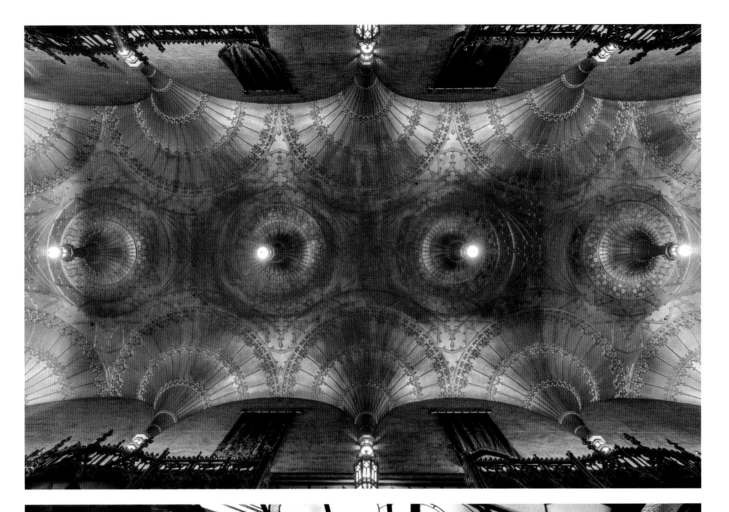

天花板

FLOOR PATTERN
地面图案

几何图形地面铺设的地面虽然看似凌乱，却有一种无法言说的沉稳，同时和墙面上的大理石花纹也有异曲同工之妙。用各种几何图形打造出极具空间感的公共空间，金属色调营造华美绚烂的视觉印象。

地面图案

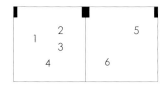

1. Mosaico Galerie Vivenne, Paris.
巴黎 Mosaico Vivienne 画廊。

2. Ministry of Education, Warsaw.
华沙教育部。

3. Lakefront Airport.
新奥尔良湖畔机场。

4. Public Service Building.
公共服务楼。

5. Pattern.
图案。

6. Pergamon Museum.
贝加蒙博物馆。

FLOOR PATTERN

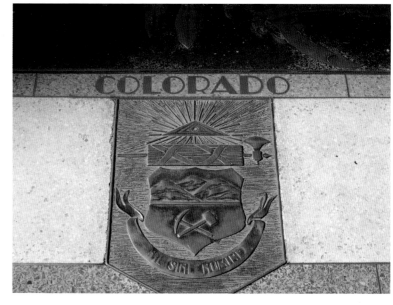

地面图案

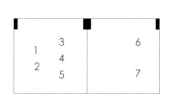

1/2/3/4/5. Hoover Dam.
胡佛水坝。

6. Miami Art Deco.
迈阿密装饰艺术。

7. Art Nouveau features in the Central Arcade built in 1906 in Newcastle Upon Tyne, UK. Designed by Harold and Joseph Oswald.
装饰艺术始于1906年的英国泰恩河的中央商场，由哈罗德和约瑟夫·奥斯瓦尔德设计。

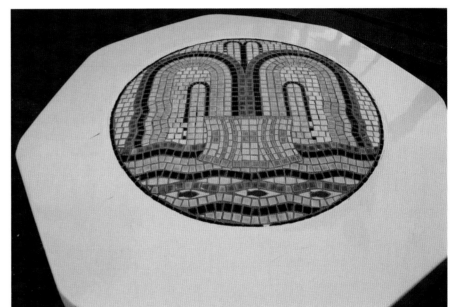

地面图案

1/2/3/4/5/6/7. Bollard with ceramic scene inset.
波兰德陶瓷场景插图。

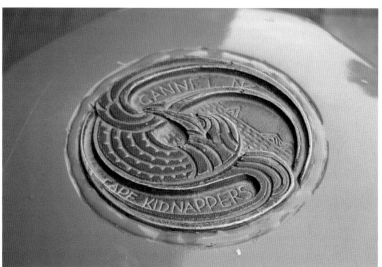

1/2/3/4/5. Bollard with metal inset.
波兰德镶嵌金属。

6/7. The Guggenheim Museums.
古根海姆博物馆。

8. Art Nouveau arcade built c.1900 by architect Frank Matcham.
装饰风艺术的骑楼建于公元前 1900 年，由建筑师弗兰克·麦青设计。

9. Quinta Avenida, Manhattan, New York.
纽约曼哈顿金塔大道。

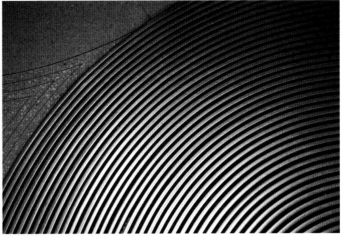

地面图案

FLOOR PATTERN

1	2	4	5
		6	7 8
	3		

1. Art Nouveau arcade built c.1900 by architect Frank Matcham.

装饰风艺术的骑楼建于公元前 1900 年，由建筑师弗兰克·麦青设计。

2. Art Nouveau features in the Central Arcade built in 1906 in Newcastle Upon Tyne, UK. Designed by Harold and Joseph Oswald.

装饰风艺术始于 1906 年的英国泰恩河的中央商场，由哈罗德和约瑟夫·奥斯瓦尔德设计。

3. Higgins Building Entrance.

希金斯大楼入口。

4/5/6/7/8. Hoover Dam.

胡佛水坝。

地面图案

FLOOR PATTERN

地面图案

1. Art deco pattern.
装饰艺术风格图案。

2. Lion d'Or, Terrazzo Floor.
水磨石地面。

3. Casa d'Italia.
卡萨科特迪瓦。

4/5/6. Havana Art Deco.
哈瓦那装饰艺术。

FLOOR PATTERN

1/2/3/4. Terrazzo Floor.
水磨石地面。

5. Severance Hall.
塞弗伦斯音乐厅。

6. Days Hotel.
戴斯酒店。

7. Carnegie LIbrary, Reims.
兰斯卡内基图书馆。

8. Art Deco Metal Floorgrate.
装饰艺术风格的金属地板格栅。

地面图案

1. Marine Building Vancouver.
温哥华海事大厦。

2/3. Hoover Dam.
胡佛水坝。

4. Canadian Legion No. 001.
加拿大军团第 001 号。

5. Art Deco Terrazzo Floor.
装饰艺术风格的水磨石地面。

6. Terrazzo Floor.
水磨石地面。

地面图案

FLOOR PATTERN

| 1 | 3 | 5 | 6 |
| 2 | 4 | | 7 |

1/2. Miami Art Deco.
迈阿密装饰艺术。

3/4. Terrazzo Floor.
水磨石地面。

5/6. Ministry of Education, Warsaw.
迈阿密装饰艺术。

7. Seal in West Hospital Lobby.
医院西大厅。

地面图案

地面图案

1/2/3/4/5. Milano Centrale.
米兰中央火车站。

FLOOR PATTERN

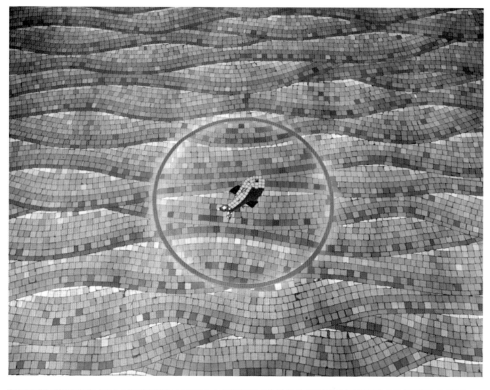

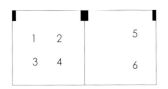

1. Rio de Janeiro Art Deco.
里约热内卢的装饰艺术。

2. Monument in Pedrosa de la Vega, Spain.
西班牙佩德罗萨德拉维加纪念性建筑。

3. Musée national des Arts d'Afrique et d'Océanie.
非洲艺术博物馆。

4. Pontefract Museum, built in 1904 by George Pennington in an Art Nouveau style as Carnegie Public Library.
庞蒂弗拉克特博物馆，由乔治潘宁顿建于1904年，作为装饰风艺术的卡内基公共图书馆。

5. Havana Art Deco.
哈瓦那装饰艺术。

6. Hoover Dam.
胡佛水坝。

地面图案

268

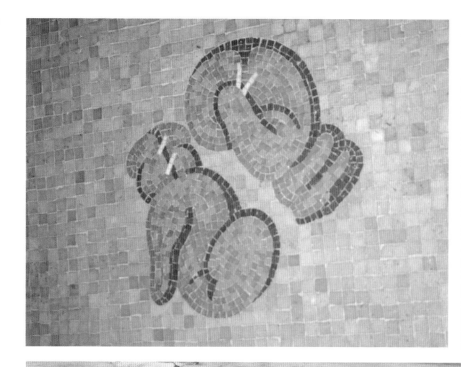

FLOOR PATTERN

地面图案

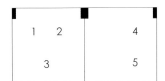

1/2/3. Rome Art Deco.
罗马装饰艺术。

4. Mosaic floor clock.
马赛克落地钟。

5. Art Deco Fireplace.
装饰艺术壁炉。

WINDOW
窗 户

将欧洲教堂彩色玻璃技法应用到窗户之上，打造出具有浓厚复古风味的窗饰装饰，彩色玻璃装饰屏风、彩色玻璃门等，玻璃元素和装饰图案的应用，也能迅速提升家中的装饰味道，提升空间品味，打造地道的 Art Deco（装饰主义）家居。

WINDOW

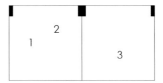

1. Beautiful Art Deco window in historical part of Luxembourg city.
那美丽的装饰艺术的窗口是卢森堡市历史遗留的一部分。

2. Ticket office, Obecní dum.
布拉格市民会馆售票处。

3. Nantes la jolie.
南特拉·朱莉。

窗户

WINDOW

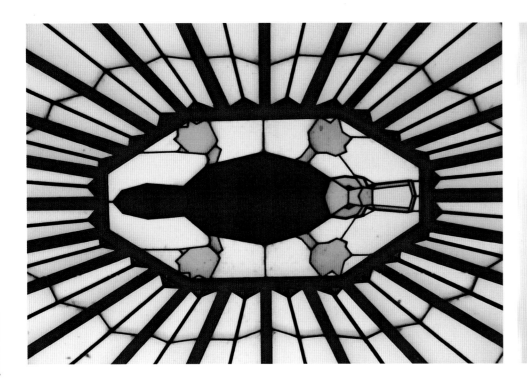

1/2. Silesian Parliament.
西里西亚议会。

3. Skylight in former Institute of Anatomy.
解剖学研究所内的天窗。

4. Stained glass decorated window above a door, this style is also known as amsterdam school.
门梁上饰有彩色玻璃的窗户，这种风格也闻名于阿姆斯特丹学校。

5. Queen's Theatre.
伦敦女王剧院。

6. Beautiful building with floral decoration and closeup window in Art Deco style(Prague, Czech Republic).
捷克共和国布拉格饰有美丽花卉图案的建筑及装饰艺术的窗户。

窗户

窗户

1. The Lis is a modernist palace home of Salamanca built on the former stretch of the wall of the city, where the Museum of Art Nouveau and Art Deco works.

李斯大厦是萨拉曼卡建在城市外围墙上的现代宫殿，以前是新艺术和装饰艺术博物馆。

2/3. Art Deco Montevideo.

装饰艺术风格的窗户。

4. Art Deco Montevideo.

乌拉圭首都蒙得维的亚装饰艺术风格的室内。

WINDOW

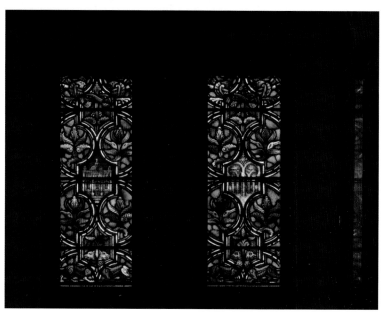

1/2/3. Saint Francis Xavier Catholic Church Stained Glass Window, New York City.
纽约圣弗朗西斯泽维尔天主教教堂内的玻璃窗户。

4. Temple Emanu-El Synagogue Stained Glass Windows, New York City.
纽约寺伊曼纽尔·萨尔瓦多犹太教堂内玻璃窗户。

5. Tower of London, Window.
伦敦塔内的窗户。

6. Av. da Liberdade, Lisbon, Portugal.
葡萄牙里斯本自由大道。

7. Window of Roosevelt University.
罗斯福大学内的窗户。

窗户

WINDOW

1. The famous Galleria Vittorio Emanuele II in Milan, Italy, with special attention for the ceiling of an almost golden gate to luxury.
著名意大利米兰维托里奥·埃马努埃莱二世长廊内引人注目而奢华壮观的金黄天花板。

2. 570 Lexington Ave.
纽约莱克星顿大道570号。

3. A detail of the clock in the railway station of Cais do Sodre in Lisbon, Portugal.
葡萄牙里斯本卡伊斯多索德雷火车站内的钟表特写。

4. Airport, Washington DC.
美国华盛顿机场。

5. Prague Art Deco.
布拉格装饰艺术。

窗户

WINDOW

1/2/3/7. Meyer May House, Grand Rapids Interior Stained Glass.
大急流城迈耶月楼的内饰彩色玻璃。

4. Bournville Baths - windows.
伯恩维尔浴场的窗户。

5. Stained glass - Light on stone P1070474.
浅石彩色玻璃。

6. Staircase Roosevelt University.
罗斯福大学内楼梯。

8. Flinders Street Station, Melbourne.
墨尔本弗林德斯街火车站。

窗户

STAIRS
楼 梯

优美浪漫的圆舞曲,楼梯间奏出的华丽乐章。

在米色石材柱头上精心雕刻着 Art Deco（装饰主义）几何元素的图案，底座是坚实的黑白花纹的基座。石材地面上对称的几何图案的设计搭配精致优雅的楼梯，流淌的线条演绎 Art Deco（装饰主义）风格，流畅的黄铜扶手、黑色的侧边装饰，以及末端漂亮的收尾造型，如花蔓般绽放的有机线条令 Art Deco（装饰主义）的柔美贯穿了整个空间。

1. Washington DC internal capitol dome view.
美国华盛顿国会大厦室内穹顶。

2. A historic building with staircase.
一座古老大厦内的楼梯。

3. Antique stairway in Bank Indonesia, Jakarta, Indonesia.
印度尼西亚雅加达银行内历史悠久的阶梯。

4/5. The famous Galleria Vittorio Emanuele II in Milan, Italy, with special attention for the ceiling of an almost golden gate to luxury.
著名意大利米兰维托里奥·埃马努埃莱二世长廊内引人注目而奢华壮观的金黄天花板。

楼梯

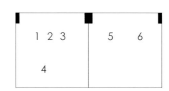

1. Art Deco Lobby.
装饰艺术风格大堂。

2. Staircases in the entrance of the Old Mississippi State Capitol building in Jackson, Mississippi.
密西西比州杰克逊市的前密西西比州议会大厦内入口的楼梯。

3. Airport, Washington DC.
美国华盛顿机场。

4. Marble staircase
大理石楼梯。

5. The Hoover Building-The Main Stairwell, The Hoover Building (1931-1935) by Wallis Gilbert and Partners, Western Avenue, London.
伦敦西大街胡佛大楼的主楼梯间。胡佛大厦（1931~1935），由沃利斯·吉尔伯特建筑公司设计。

6. Tokyo Metropolitan Teien Art Museum.
东京都庭园美术馆。

楼梯

1/2. Burbank City Hall.
伯班克市市政厅。

3. Carbide and Carbon Building Chicago.
芝加哥联碳大楼。

3. National Breweries.
国家啤酒厂。

4. Stairs in GE Building, Rockerfeller Centre.
洛克菲勒中心奇异电器大楼内的楼梯。

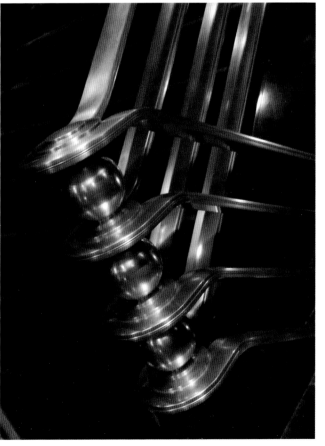

楼梯

STAIRS

1. Chicago Cultural Center.
芝加哥文化中心。

2. Curves, Contours and Arches.
曲线，轮廓和拱门。

3/4. Grand Ocean Hotel, Saltdean. Ex Butlins holiday camp, now converted to luxury flats. The four main accommodation wings have been demolished, but the main block has been retained.
大洋酒店，前身是布特林度假夏令营，现在被改造成了奢华公寓。四个主要的住宿区已经拆除，但是主大厦依旧保留着。

5. Carnegie Library, Reims.
兰斯卡内基图书馆。

楼梯

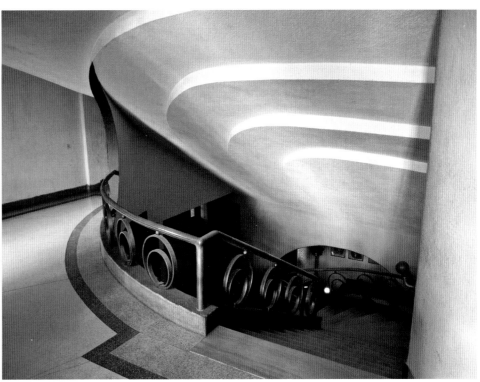

楼梯

| 1 | 2 | | 5 | 6 |
| 3 | 4 | | | 7 |

1. QVB Dome & Escalators.
维多利亚女王大厦的穹顶及扶梯。

2. Palais de Justice.
司法宫。

3. Ascending staircase.
向上延伸的楼梯。

4. Queen's Theater.
伦敦女王剧院。

5/6/7. Havana Art Deco.
哈瓦那装饰艺术。

STAIRS

楼梯

1. Greyhound terminal.
洛杉矶"灰狗"汽车总站。

2. Conseil d'Etat.
行政法院。

3/4/5/6. Public Service Building.
公共服务大厦。

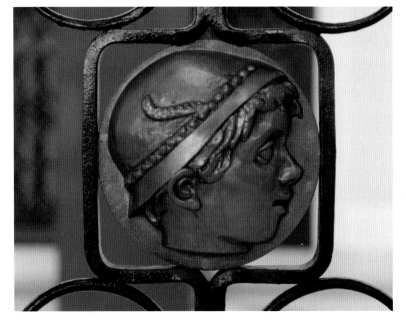

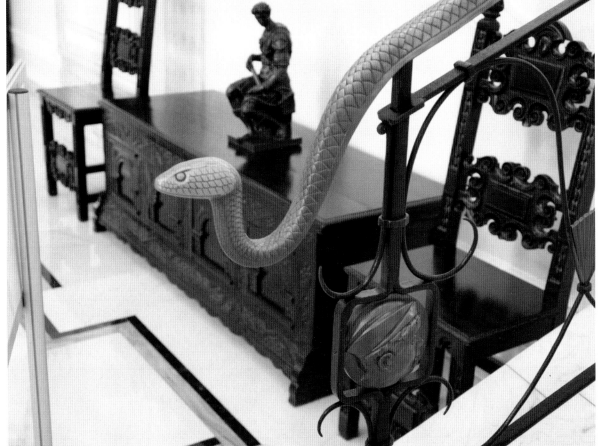

1/2/3/4/5. Parliament of Poland.
波兰议会厅。

楼梯

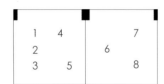

1/2/3/4. Queen's Theater.
伦敦女王剧院。

5. Standard Building 1924.
美国标准大厦，建于 1924 年。

6. Horta Museum.
奥塔博物馆。

7. Beautiful Art Nouveau staircase in the Klothild Palace, Budapest. The palace is now the Buddha-Bar Hotel.
布达佩斯克洛丝尔德宫殿内美轮美奂的新艺术风格楼梯，现是佛陀酒吧巴黎酒店。

8. Marble floor with stainless steel staircase.
配有大理石地面的不锈钢楼梯。

楼梯

1. At Hotel Pariz in Prague.
布拉格巴黎大酒店。

2. Art Nouvau staircase in a building in Bruxelles, Belgique.
比利时布鲁塞尔一座大厦内的楼梯。

3/4/5/6. Labour Exchange.
职业介绍所。

楼梯

1/2. Grand lobby in Cunard's Queen Mary.
维多利亚女王号邮轮内奢华大堂。

3. Napier one of the cities with the best collection of Art Deco buildings in the world.
纳皮尔是世界上集结装饰艺术风格建筑最佳的城市之一。

楼梯

1/2. QVB Stairway.
维多利亚女王大厦内阶梯。

3. The staircase at the Grand Palais.
巴黎大皇宫美术馆内的楼梯。

4/5. Public Service Building.
公共服务大厦。

楼梯

CLOCK
钟 表

运用辐射状的灯光、对称简洁的几何构图以及放射性罗马数字于钟表表盘上,运输工具如火车、汽车和很多海报中都可见其身影。钢架结构也是主要的表现题材,以埃菲尔铁塔为钢架设计的代表。粉色、暗红、叠加后的蓝色、金属味的银白和古铜色都被运用于大部分 Art Deco(装饰主义)风格钟表设计中。

在恒久的时光节奏中,不徐不疾地摆动着的钟摆,躲进 Art Deco(装饰主义)里的挂钟。

CLOCK

钟表

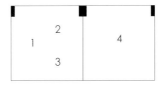

1. Art deco clock at Fleet street in London.
伦敦舰队街装饰艺术风格的钟表。

2. The centre of British finance, the clock on the Royal Exchange building and the Bank of England to the right.
处于英国的金融中心皇家交易所大楼与英国央行交汇处右边的时钟。

3. Golden clock at the street.
金黄色街边时钟。

4. Basel Art Deco Clock.
巴塞尔装饰艺术风格的钟表。

1	2	4	5
		6	8
3		7	

1. Union Terminal's clock.
辛辛那提联合车站内的时钟。

2. Clock in the boardroom of a fully decorated art deco building, a former brewery office in Breda, the Netherlands, with stained glass in the wall.
饰有墙上玻璃的装饰艺术风格大楼,之前是荷兰布雷达啤酒厂办公室。

3. Portland stone is a limestone of Portland, Dorset.
波特兰石是多塞特波特兰岛的石灰石。

4. Art Deco Clock - South Boston Station.
波士顿南站内装饰艺术风格的时钟。

5/6. St. Olaf House-Clock, St. Olaf House, Hay's Wharf (1929-31) by H.S. Goodhart-Rendel.
干草码头圣奥拉夫楼圣奥拉夫众议院内的时钟(1929年~1931年)。由前任牛津大学美术教授斯莱德.古德哈特－林德宪设计。

7. Tulsa Treasures.
塔尔萨瑰宝大厦。

8. Information service clock in Grand Central Station Hall.
纽约中央车站内服务信息台的时钟。

钟表

1. The big clock in Antwerp Central Train railway station.
安特卫普中央火车站内大钟。

2. Old astonomical clock in Riga, the capital of Latvia.
拉脱维亚首都里加历史悠久的大钟。

3. New York Art Deco.
纽约装饰艺术。

4. Guardian Building Monel metal screen.
嘉德大厦内蒙乃尔合金金属屏。

钟表

CLOCK

钟表

1. The original 1930s entrance hall to the Royal Shakespeare Theatre. The new build has kept some of the original Art Deco features.
通往20世纪30年代皇家莎士比亚剧院的入口大厅，新建筑保留了一些原有的装饰艺术特色。

2. Durham Museum, Omaha NE.
奥马哈达勒姆博物馆。

3. Art Deco clock over the entry doors in the lobby of the Gulf Tower in Pittsburgh.
匹兹堡海湾塔大厅入口处的装饰艺术风格时钟。

4. Daily Express Art Deco.
每日快报大厦内装饰艺术。

5. Art Deco clock at the Ross building.
罗斯大厦内装饰艺术时钟。

6. 570 Lexington Ave Art Deco.
纽约莱克星顿大道570号装饰艺术。

CLOCK

1	2		5	
	3	4	6	
			7	

1/2/3. Prague Astronomical Clock (Prague Orloj) - on the wall of Old Town City Hall in the Old Town Square in Prague in the Czech Republic.

布拉格天文钟——镶嵌在捷克共和国布拉格老城广场的老城镇市政厅墙上。

4. The golden clock on the wall.

墙上金色的老时钟。

5. De la Warr Pavilion.

德拉沃尔馆。

6. Grand Ocean Hotel, Saltdean.

大洋酒店。

7. Saint-Quentin Railway Station.

圣康坦火车站。

钟表

CLOCK

1. The vault at the Chicago Board of Trade opened to the public for the first time in 83 years this past weekend.
芝加哥贸易委员会的金库在过去的 83 年首次对外开放。

2. Pennsylvania Railroad Station.
宾夕法尼亚火车站。

3. Newark, Art Deco.
纽瓦克装饰艺术。

4/5. Vienna Curious Art Deco clock.
维也纳装饰艺术风格的时钟。

钟表

LIGHTING

灯具

似剪纸造型和带有几何图形的灯具,成为 Art Deco(装饰主义)空间中最深刻的部分。带有花纹的壁纸和移门,简洁的长发女白描的挂画,处处体现浪漫的情节。厨柜、书柜、沙发均采用激活空间的亮丽品蓝。平衡与折中成为 Art Deco(装饰主义)把脉的要点,柔化极简空间中机械线条,丰富冷淡色彩。空间因为 Art Deco(装饰主义)有了更多的灵气和浪漫。

地面的黑白经典 Art Deco(装饰主义)几何图案,让空间兼具经典的和摩登的特性。大堂上空极具特色的金色艺术玻璃管灯具成为空间的点睛之笔。

吊灯

在上世纪20年代,上海便逐渐兴起了Art Deco(装饰主义)建筑风格,最早的便是1923年的汇丰银行大楼,在它古典风格的大厅中,饰有浓郁Art Deco(装饰主义)风格的吊灯。

吊灯的设计灵感来源于纽约的廉价市场售卖的灯具,经过精心设计后呈现出华丽的外表和优美的线条。

灯具

1. Light shining through an art deco light fixture.
一缕光透过那装饰艺术风格的灯具。

2. Ceiling decorated with flower pattern glass ceiling.
饰有花纹玻璃的天花板。

3. Beautiful and elaborate hanging chandelier.
美丽而精致的挂吊灯。

4. Ceiling Of Art Deco Post Office South Beach-Canon Sure Shot Telemax camera and Kodak HD 400 film.
南滩邮局内装饰艺术风格的天花板。

5. Art Deco Chandelier.
装饰艺术风格的吊灯。

6. Havana Art Deco.
哈瓦那装饰艺术。

7. Art Deco Light Fixture-Lobby of the Koppers Building in Pittsburgh.
匹兹堡考伯斯大厅内装饰艺术风格的灯具。

1. Asheville City Hall.
阿什维尔市政厅。

2. Art Deco Imperial Hotel, Prague.
布拉格帝国酒店装饰艺术。

3. The Brown hotel.
布朗酒店。

4. Durham Museum, Omaha NE.
美国内华达州奥马哈达勒姆博物馆。

灯具

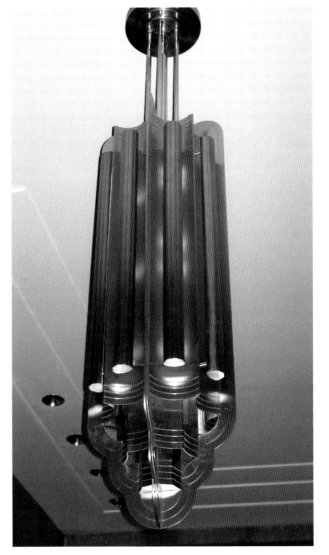

1. Deco chandelier New Yorker Hotel.
纽约客酒店内的装饰吊灯。

2. Durham Museum, Omaha NE.
美国内华达州奥马哈达勒姆博物馆。

3. Paramount Theater.
派拉蒙剧院。

4. Silesian Parliament.
西里西亚议会。

5. Vintage art deco ceiling lamp.
复古的装饰艺术吊灯。

6. Cleveland Art Deco.
克里夫兰装饰艺术。

7. Ministry of Education, Warsaw.
华沙教育部。

灯具

1. Tulsa Treasures.
塔尔萨瑰宝大厦。

2. Paramount Theater.
派拉蒙剧院。

3. lights above theatre seats.
灯光散在剧院座位上。

4. Cleveland Art Deco.
克里夫兰装饰艺术。

5. Tokyo Metropolitan Teien Art Museum.
东京都庭园美术馆。

6. Art Deco Light Fixtures - Cafe Americain - Amsterdam.
阿姆斯特丹美国咖啡屋内的装饰艺术风格灯具。

7. Washington State Capitol Interior Dome and Chandelier.
华盛顿国会大厦内穹顶及吊灯。

8. The Frist Center for the Visual Arts.
弗里斯特视觉艺术中心。

灯具

LIGHTING

灯具

1		4	5	6
2	3	7	8	

1. Public Service Building.
公共服务大厦。

2. The Art Deco central chandelier (lampe du théâtre), Théâtre des Champs Elysees, Paris.
巴黎香榭丽舍大街剧院内装饰艺术风格的吊灯

3. Chandeliers inside the historic palace
一座历史悠久宫殿内的吊灯。

4. Inside the Kirby Center.
内柯比中心。

5/6. Art Nouveau lights, Obecní dum, Prague.
布拉格市民会馆内新艺术风格的吊灯。

7. Ministry of Education, Warsaw.
华沙教育部。

8. Clock in arcade below our apartment.
我们公寓下的拱廊时钟。

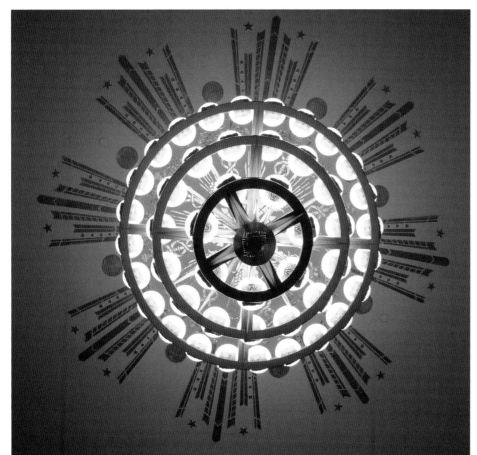

灯具

1. Queen Mary Hotel, Long Beach.
长滩玛丽皇后酒店。

2. Chandelier in Hotel
酒店内的挂灯。

3. Newark, Art Deco.
纽瓦克装饰艺术。

4. Art deco chandelier
装饰艺术风格的吊灯。

5. Tulsa Treasures.
塔尔萨瑰宝大厦。

6. Salon of Light.
美发店内的吊灯。

灯具

1. 30th Street Station, Philadelphia.
美国费城第三十街站。

2. Art deco restaurant.
装饰艺术风格餐厅。

3. Municipal House in Prague.
布拉格市民会馆。

4. Havana art deco.
哈瓦那装饰艺术。

落地灯

Art Deco（装饰主义）风格的表现，典型代表之一，即是线条。无论是放射状的装饰镜面或者落地灯，整齐重复的锯齿纹路等都是 Art Deco（装饰主义）风格的重要表现形式。金和棕色的搭配，也是打造奢华家居的重要手法，棕色的大气稳重，经过了历史的洗练，霸气的姿态，王者的盛宴。

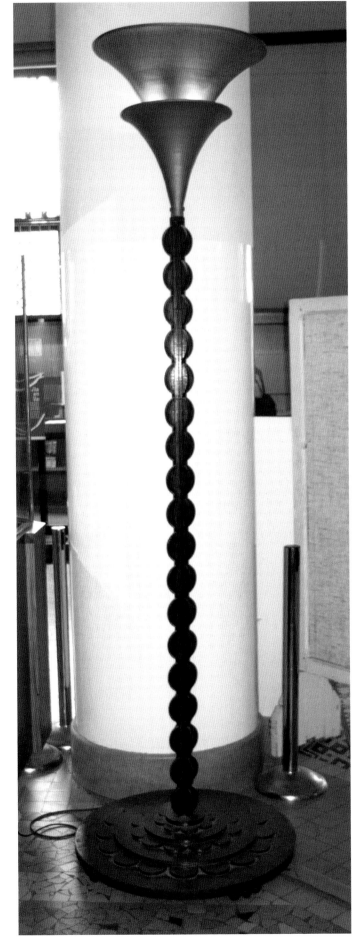

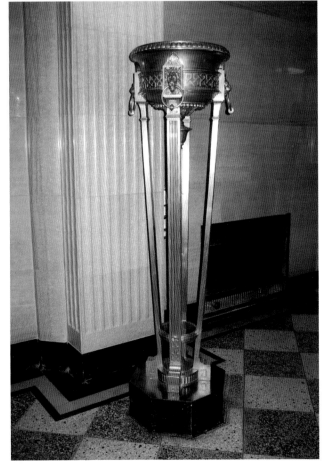

1. Church of the Holy Spirit.
圣灵大教堂。

2. Art Nouveau inspired escalators in Harrods, Brompton Road, London.
伦敦布朗普顿路哈罗德新艺术运动风格的扶梯。

3. National Museum of Arts of Africa and Oceania.
非洲和大洋洲艺术国家博物馆。

4. Court of Appeal Montreal.
蒙特利尔上诉法院。

5. Dominion Government Building.
自治领政府大楼。

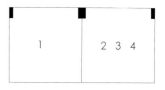

1. Severance Hall.
塞弗伦斯音乐厅。

2. Courthouse Annex (Court of Appeal).
法院（上诉法院）附件。

3. Torchiere.
炬灯。

4. Church of the Holy Spirit.
圣灵大教堂。

灯具

壁灯

Art Deco（装饰主义）风格的照明以辅助光源为主，更强调壁灯设计，更强调温馨、高雅的格调。Art Deco（装饰主义）风格的照明和饰品结合讲究的是布局的审美效果跟整体的格调统一的固定样式。

1. Metal lamp, glass block window and a green wall in the art deco style. Shot in Rockport, Texas.
装饰艺术风格的金属灯，玻璃窗口及绿墙，摄于得克萨斯州罗克波特。

2. Lecce.
意大利莱切。

3. Cleveland Art Deco.
克里夫兰装饰艺术。

4. Paris Art Deco.
巴黎装饰艺术。

5. Ministry of Education, Warsaw.
华沙教育部。

1 2 3		6
	5	
4		7

1. Durham Museum, Omaha NE.
奥马哈达勒姆博物馆。

2. Wall lamp.
壁灯。

3. Barber Institute of Fine Arts - lamps.
美术理发院内的灯。

4. Lamp at Carew Tower.
卡鲁塔内的灯。

5. Carbide and Carbon Building Chicago.
芝加哥联碳大楼。

6. Bellas Artes.
美术宫。

7. Deco lamp.
装饰灯。

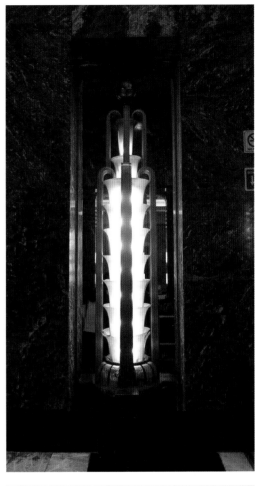
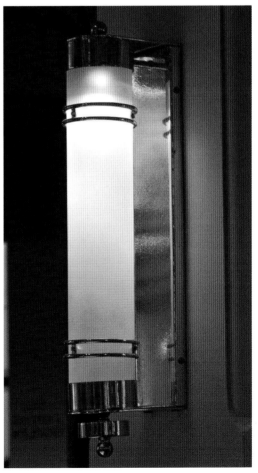

灯具

灯具

1/2. Havana Art Deco.
哈瓦那装饰艺术。

3. Aldred Building.
奥尔德雷德大楼。

4. Cleveland Art Deco.
克里夫兰装饰艺术。

5/6. Iron works.
铁艺作品。

7. Light Fixture in Carew Tower Arcade.
卡鲁塔商场内的灯具。

8. Silesian Parliament.
西里西亚议会。

1. Art Deco designed doors and light fixtures in a Chicago building.
芝加哥一座大楼内装饰艺术风格设计的门和灯。

2/3. Durham Museum, Omaha NE.
奥马哈达勒姆博物馆。

4. The Carlu.
木·卡路。

5. Newark Art Deco.
纽瓦克装饰艺术。

6. A fully decorated art deco building, a former brewery office in Breda, the Netherlands, with stained glass in the wall.
饰有墙上玻璃的装饰艺术风格大楼，之前是荷兰布雷达啤酒厂办公室。

7. Court of Appeal Montreal.
蒙特利尔上诉法院。

8. Silesian Parliament.
西里西亚议会。

灯具

WAYFINDING & SIGNAGE
导视标识

Art Deco（装饰主义）大量运用了鲨鱼纹、斑马纹、曲折锯齿图形、阶梯图形、粗体与弯曲的曲线、放射状图样的字体雕刻在指示牌、导视设计和标识上。

Art Deco（装饰主义）的无限魅力，就在于对装饰淋漓尽致的运用，且不论时代如何变迁，都能在其中出现新突破。装饰艺术运动可以说是整个设计界的探索，由于在各国的表现形式融合了当地的本土特征，更加多元化，所以很难在世界范围内形成统一、流行的风格。但它仍具有一致的特征，如注重表现标识材料的质感、光泽；造型设计中多采用几何形状或用折线的形状进行装饰；导视系统的色彩设计中强调运用鲜艳的纯色、对比色和金属色，造成华美绚烂的视觉印象。

1. The Barber Institute of Fine Arts - plaque - Architecture Medal - 1946 from the R.I.B.A.
美术理发院内英国皇家建筑师学会建筑奖章。

2. Storm Drain Cover.
雨水管盖。

3. Traditional shop sign in Buda, Budapest, Hungary.
匈牙利布达佩斯布达的传统店铺招牌。

4. Portland stone is a limestone of Portland, Dorset.
波特兰石是多塞特波特兰岛的石灰石。

5/6. The Barber Institute of Fine Arts - coat of arms.
美术理发院内纹章。

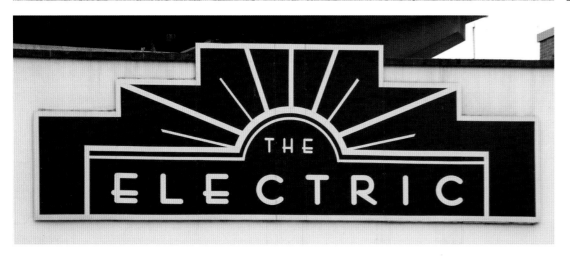

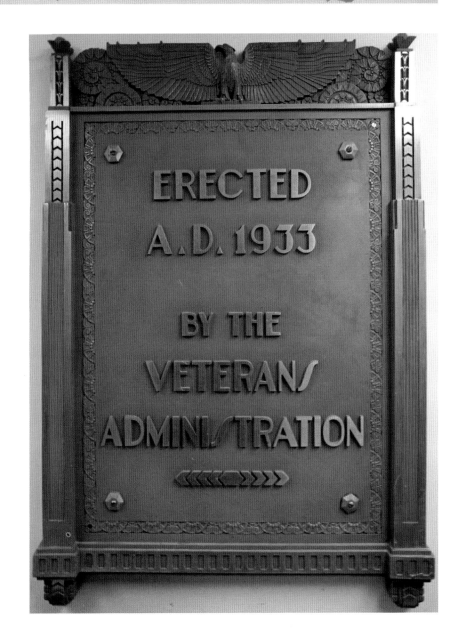

1/2. A wonderful 1930s Art-Deco school in Quimperle
坎佩莱20世纪30年代美轮美奂的装饰艺术风格学校。

3. The Electric Cinema opened in Station Street in 1909, showing its first silent film on 27 December of that year, and is now the oldest working cinema in the country.
电动电影院于1909年在车站街开业，那年12月27日上映了第一部哑剧，它是现在最古老的电影院。

4. World Premiere Lendl Mucha exhibition at the Obecní dum.
布拉格市民会馆伦德尔木栅展览世界首演。

5. Daily Express Art Deco.
每日快报大厦内的装饰艺术。

6. Building date.
建造日期。

7. San Francisco VA Medical Center (SFVAMC).
旧金山医学中心。

导视标识

WAYFINDING & SIGNAGE

1. Library signage.
图书馆标识。

2. 35 East Wacker Drive (Jewelers Building).
芝加哥东瓦克车道 35 号大楼（珠宝大楼）。

3. Mine was called Phoenix.
被称为凤凰的煤矿。

4. The Roosevelt.
罗斯福。

5. White House.
白宫。

6. Hearst Building, San Francisco, California.
旧金山加利福尼亚赫斯特大厦。

7. Marine Building Vancouver.
温哥华海事大厦。

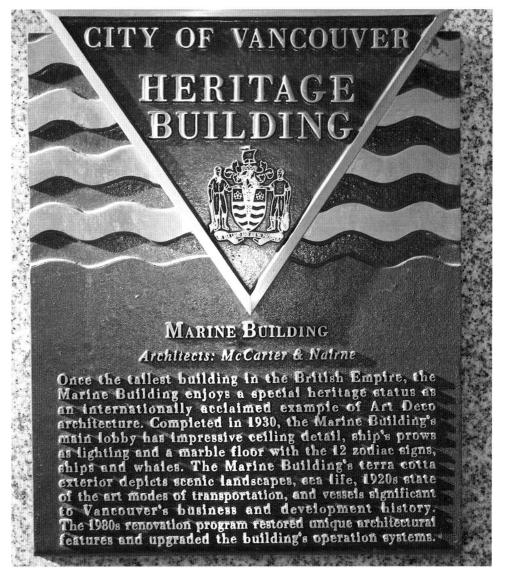

导视标识

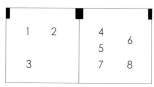

1. Downtown Miami Art Deco and 1920's buildings.
迈阿密市中心的1920年装饰艺术风格建筑。

2. Door Handle.
门把手。

3. Empire State Building.
美国帝国大厦。

4/5/6. Firenze Santa Maria Novello station, built in 1932.
佛罗伦萨的圣塔玛丽亚诺韦洛站建立于1932年。

7/8. Elevator Sign DuPont Building Downtown Miami.
迈阿密市中心杜邦大厦内的电梯标志。

导视标识

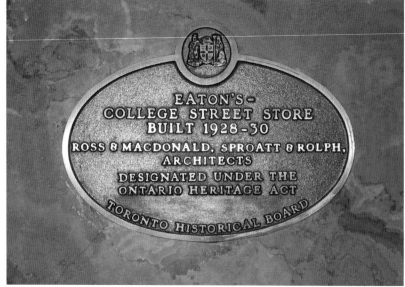

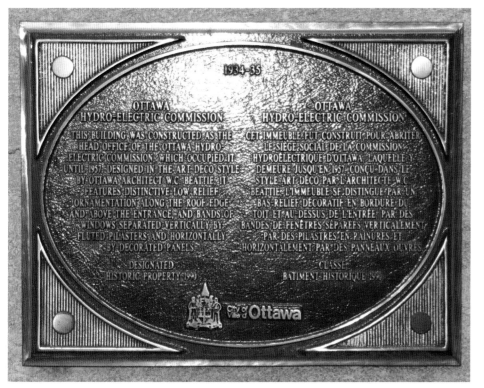
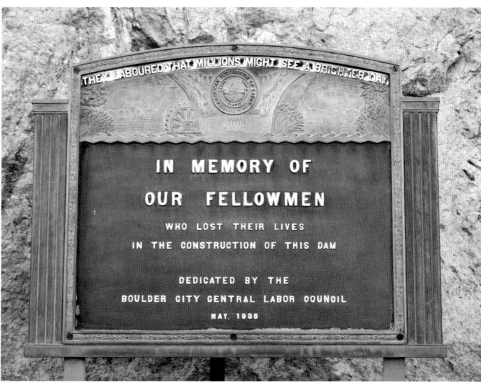

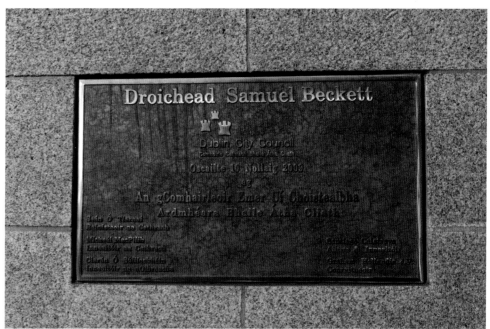

导视标识

1. Signage of Walter Percy Chrysler's
沃尔特·佩尔西克莱斯勒标识。

2. Signage of Eaton's College Street Store.
伊顿学院街店标识。

3. Standard Building 1924.
标准大厦建于1924年。

4. Croydon Airport, the worlds first purpose built air teminal. Built in 1928, and now lovingly looked after.
克罗伊登机场是世界第一个特制航空终点站，建于1928年，现被妥善管理。

5. Carnegie LIbrary, Reims.
兰斯卡内基图书馆。

6. Supreme Court.
最高法院。

7. Hoover Dam.
胡佛水坝。

8. Brass sign plaque outside flagship Macy's Department Store at Herald Square in NYC. Macy's was founded by Rowland Hussey Macy in 1851.
纽约先驱广场梅西旗店百货外的铜招牌牌匾。梅西旗舰店百货在1851年由罗兰赫西梅西创立。

9. Dublin Docklands - Samuel Beckett Bridge.
都柏林码头区——贝克特塞缪尔大桥。

1/2. Empire State Building Interior Art Deco Motif.
帝国大厦室内装饰艺术图案。

3. Rising Atlanta.
亚特兰大的崛起。

4. Los Angeles Union Station.
洛杉矶联合车站。

5. Museum of London.
伦敦博物馆。

6. 4 Av-9 St Station.
第四大道第九街站。

导视标识

WAYFINDING & SIGNAGE

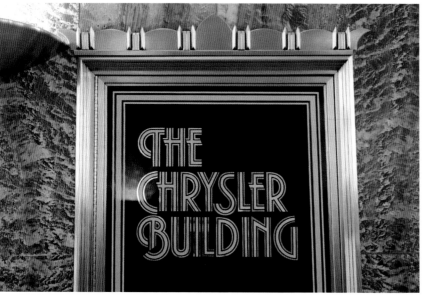

1. Metro sign.
地铁标志。

2. Chrysler Building.
克莱斯勒大厦。

3. The Alexandra Theatre on John Bright Street – 1935 rebuilding – sign – Main entrance on Suffolk Street.
1935年重建的约翰布莱特街亚历山德拉剧院内的标识——主入口在萨福克街。

4. Sign and the Legion of Honor at Tapestry Museum in Aix-en-Provence.
普罗旺斯地区艾克斯博物馆内标识和荣誉勋章。

5/6/7. Historic 9th District Court of Appeals.
历史第九区上诉法院。

8/9. Art Deco.
装饰艺术。

导视标识

FURNITURE & FURNISHINGS
家具摆设

奢华、高贵的铁艺，带有质感的华丽气氛萦绕。金属味的茶几更是用简单的线条营造尊贵的现代感。支撑桌子的桌体看似繁乱，带有机械的科技感和现代感。机械式的格纹和钢铁材质结合，缔造了不同凡响的入户方式，将户外的自然风光与室内优雅的人文情怀进行链接，展示 Art Deco（装饰主义）风格的感性和韵味。质感十足的家具，从每一个细节处体现 Art Deco（装饰主义）的风格特色。

家具的精致，点睛质感的空间。

FURNITURE & FURNISHINGS

家具摆设

1. New York Art Deco.
纽约装饰艺术。

2. Saint-Jean-Berchmans.
圣让贝克曼斯。

3. OPORTO. Casa Condes de Vizela. Hall con consola.
波尔图卡萨维泽拉堂庭院内控制台。

4/5. Paramount Theater.
派拉蒙剧院。

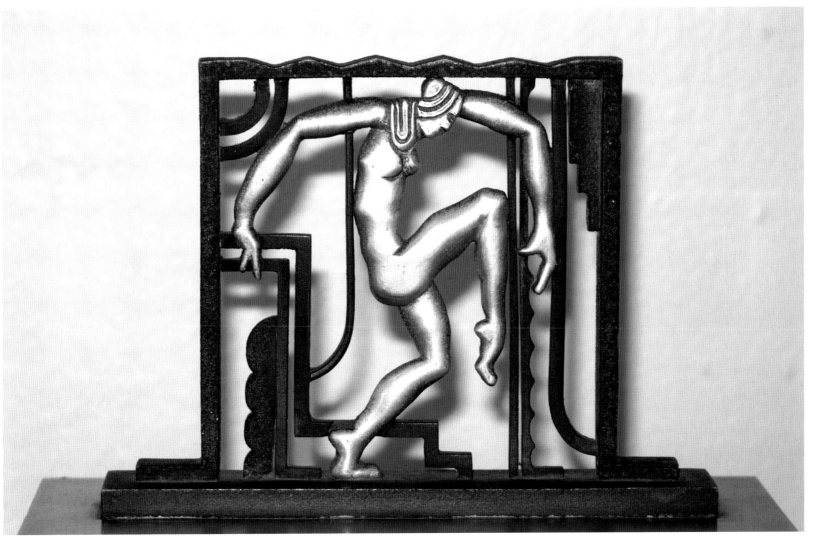

家具摆设

1. Lakefront Airport.
新奥尔良湖畔机场。

2/3/4/5. Iron works.
铁艺作品。

1. Oviatt penthouse.
奥维亚特阁楼。

2/3/4. Iron Works.
铁艺。

5. Saint-Jean-Berchmans.
圣让贝克曼斯。

家具摆设

CORRIDOR SPACE
过道空间

Art Deco（装饰主义）风格常被注入到酒店内的各个过道角落，为住客掀起一场装饰艺术的盛宴。由充满装饰风格的地毯和壁纸组成的长廊，这种设计在客房里更是到了夸张的地步，从地毯、床头垫、落地灯到空调出风口等都充满了 Art Deco（装饰主义）的细节，过道墙壁上的挂画是装饰艺术建筑的黑白老照片。

CORRIDOR SPACE

1. Cleveland Art Deco.
克利夫兰装饰艺术。

2. Court of Appeal Montreal.
蒙特利尔上诉法院。

3. An elevator lobby with carpet and modern lighting.
设有地毯和现代灯具的大堂电梯。

4. Art Deco-style wrought iron gate.
装饰艺术风格的铁艺门。

5. Courthouse: portrait corridor.
法院设有肖像的走廊。

过道空间

CORRIDOR SPACE

1. Balcony Entrance to Hall of Mirrors.
通往镜厅的阳台入口。

2. Fort Worth Texas post office.
沃斯堡得克萨斯州邮局。

3/4. Memorial Hall in the lobby of the Louisiana State Capitol building in Baton Rouge, Louisiana.
路易斯安那州巴吞鲁日国会大厦纪念馆的大厅。

5. Guardian Building.
嘉德大厦。

过道空间

CORRIDOR SPACE

1. Fox Wilshire Theater.
福克斯威尔希尔剧院。

2. Havana Art Deco.
哈瓦那装饰艺术。

3. Art Deco elevator lobby.
带扶梯装饰艺术风格的大堂。

4. Royal Mezzanine Circle.
皇家包厢。

5. Fort Worth Texas post office.
沃斯堡得克萨斯州邮局。

过道空间

CORRIDOR SPACE

1. Chrysler Building Lobby and elevators view in the main hall with its unique art deco style decoration.
从大堂一眼望去那独特装饰艺术风格的克莱斯勒大厦大堂和电梯。

2. View From the Southern Window of the Dixie Terminal.
从迪克终点站的西南窗边一眼望去。

3. The Hoover Building-Main Lobby, The Hoover Building (1931-1935) by Wallis Gilbert and Partners, Western Avenue, London.
伦敦西大街胡佛大楼的主楼梯间,胡佛大厦(1931~1935),由沃利斯·吉尔伯特建筑公司设计。

4. Miami Art Deco.
迈阿密装饰艺术。

过道空间

CORRIDOR SPACE

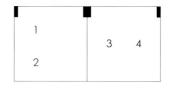

1. Guardian Building interior – retail promenade.
嘉德大厦室内——零售长廊。

2. Guardian Building – Main Lobby.
嘉德大厦——主厅。

3. Guardian Building – Lobby stepped arches.
嘉德大厦——大堂阶梯式拱门。

4. Guardian Building – Elevator Lobby.
嘉德大厦——电梯大厅。

过道空间

CORRIDOR SPACE

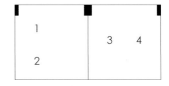

1. Palm Court.
棕榈阁。

2. Framed in Splendor.
奢华壮丽。

3. Long Art Deco corridor in hotel.
装饰艺术风格的酒店长廊。

4. Corridor at Court of Justice.
法院走廊。

过道空间

1. Terrace and pillar with window light in Ban Pun Palace/Phra Ram Ratchaniwet.
拉玛五世宫殿内的露台及窗边透光的柱子。

2. Judge Sarah T. Hughes' courtroom.
大法官莎拉·T·休斯的审判室。

3. Obecní dum.
布拉格市民会馆。

4. Lobby of Rockefeller Center (GE Building) in New York City.
纽约市洛克菲勒大厦（通用电气大厦）大堂。

5. Interior lobby of the Chrysler Building in New York, NY. The lobby is designated a national Art Deco Interior Landmark.
纽约克莱斯勒大厦的内部大厅，它被指定为国家室内的装饰艺术地标。

过道空间

CORRIDOR SPACE

1. Hilton Netherland Plaza.
希尔顿荷兰广场。

2. Carew Tower.
辛辛那提加露大楼。

3/4/5. The Petronas Towers, also known as the Petronas Twin Towers, are twin skyscrapers in Kuala Lumpur, Malaysia.
吉隆坡石油塔，又名吉隆坡石油塔，是马来西亚吉隆坡的摩天大楼。

过道空间

CORRIDOR SPACE

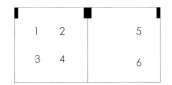

1. Interior lobby of the Chrysler Building in New York, NY. The lobby is designated a national Art Deco Interior Landmark.
纽约克莱斯勒大厦的内部大厅,它被指定为国家室内的装饰艺术地标。

2. Lobby-ist's Dream.
大堂说客的梦想。

3. Moscow Metro.
莫斯科地铁站。

4. Durham Museum, Omaha NE.
奥马哈达勒姆博物馆。

5. Belorusskaya (Koltsevaya Line).
白俄罗斯站(环转线)。

6. Senate Chamber in the Louisiana State Capitol building in Baton Rouge, Louisiana.
路易斯安那州巴吞鲁日国会大厦内的路易斯安那州参议院。

过道空间

1/2. Durham Museum, Omaha NE.
奥马哈达勒姆博物馆。

3. Art deco building
装饰艺术风格大楼。

4. The Clinton Museum.
克林顿博物馆。

5. The Frist Center for the Visual Arts.
弗里斯特中心视觉艺术。

过道空间

CORRIDOR SPACE

过道空间

1. Lobby, Southern California Edison Building.
南加尼福利亚州爱迪生大厦内的大堂。

2/3. MGM Grand.
米高梅大酒店。

1. Inside The Frist Center for the Visual Arts.
弗里斯特中心视觉艺术的室内。

2. Mayakovskaya is a Moscow metro station on the Zamoskvoretskaya Line, in the Tverskoy District of central Moscow.
马雅科夫斯基站是莫斯科河畔线上的莫斯科地铁站，位于莫斯科市中心特维尔区。

3. Cardiff Art Deco Modernism.
装饰艺术现代主义风格的加的夫。

4. Old and golden corridor at an Art Deco building.
金色古老的装饰艺术风格走廊。

过道空间

RAILING & BARRIER
护栏栏杆

唯美与时尚的艺术线条、凹凸有致的经典铁艺设计和巧夺天工的浅浮雕。Art Deco（装饰主义）被赋予高耸挺拔和复古贵族气质，尽显世界历史史上登峰造极的摩登风格。

在 Art Deco（装饰主义）风格的室内设计中，大门、栏杆常常会用金属打造出花纹、文字、图案、几何造型等。将带有此类图案的金属镂空面板作为隔断，或作为墙面装饰，就是一种地道的装饰方法。

1. Church of the Holy Spirit.
教会圣灵。

2. New York Art Deco.
纽约装饰艺术。

3. OPORTO.Fundación Serralves.Casa Condes de Vizela.Puerta.
波尔图卡萨维泽拉堂庭院内控制台。

4. Park Ave Station.
公园大道站。

5. Radiator grill.
散热器护栅。

6/7. Katowice.
卡托维兹。

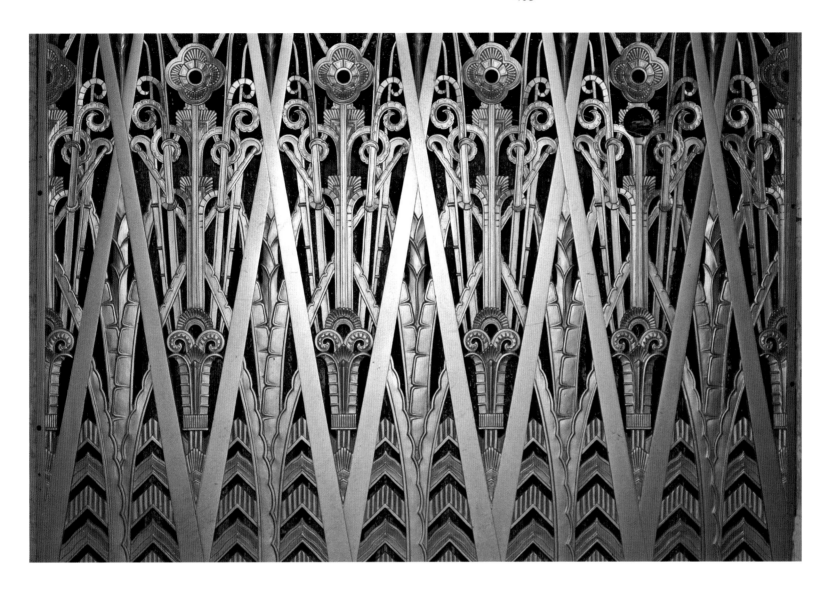

护栏栏杆

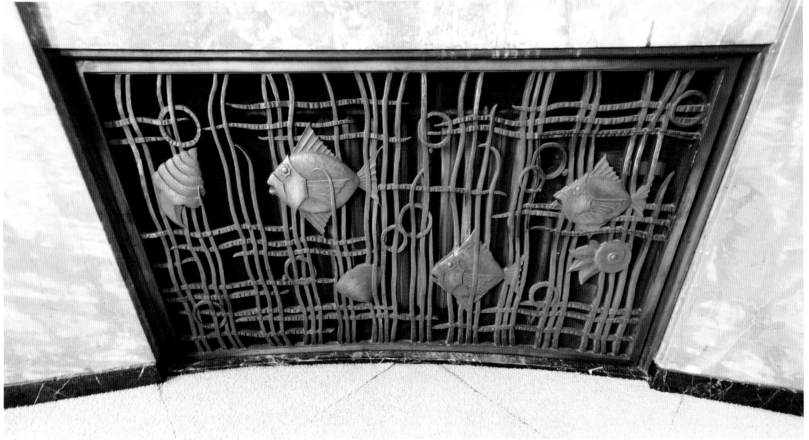

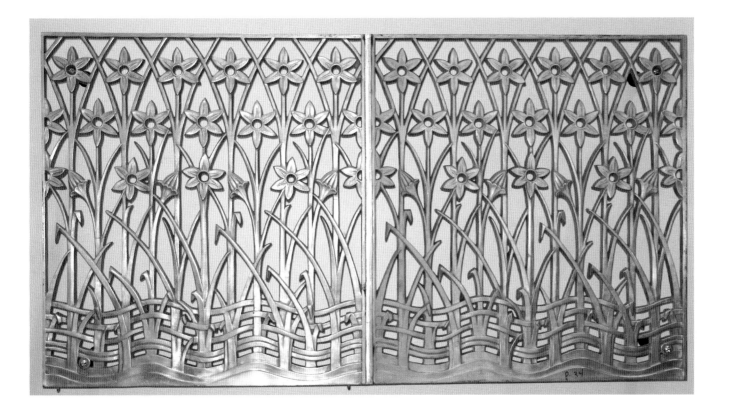
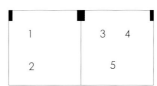

护栏栏杆

1/2. Tokyo Metropolitan Teien Art Museum.
东京都庭园美术馆。

3/4/5. Cleveland Art Deco.
克利夫兰装饰艺术。

RAILING & BARRIER

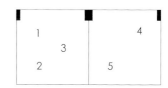

1/2/3. Marine Building Vancouver.
温哥华海事大厦。

4. Delicate wrought iron fence.
精致的铁艺护栏。

5. Standard Building 1924.
标准大厦建于1924年。

护栏栏杆

护栏栏杆

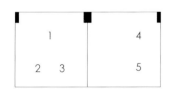

1. Art Deco Brass Railing-Decorative railing on staircase leading from lobby to basement level.
通往大堂到地下室并配有黄铜栏杆及装饰栏杆的装饰艺术风格楼梯。

2. Prague Art Deco.
布拉格装饰艺术。

3. Ottawa Hydro-Electric Commission.
渥太华水电委员会。

4. Saint-Jean-Berchmans Church.
圣让贝克曼斯教堂。

5. Marine Building Vancouver.
温哥华海事大厦。

1/2/3/4. Iron Works.
铁艺。

护栏栏杆

1/2/3/4/5/6/7/8. An Art Deco staircase added to a Victorian arcade.

配有维多利亚时代拱廊的装饰艺术风格楼梯。

护栏栏杆

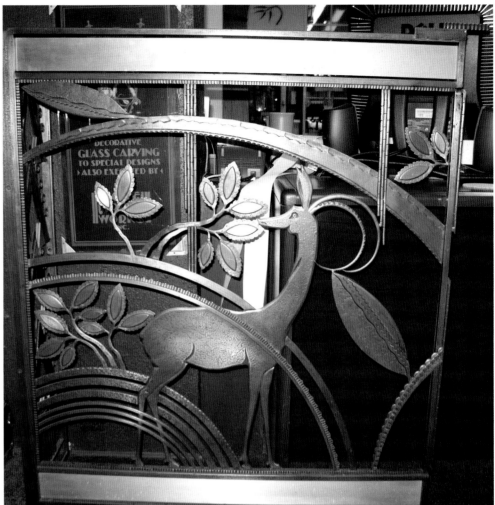

1/2/3. Iron works.
铁艺作品。

4. Features stained glass windows and wooden railings white church.
设有彩色玻璃窗户及白色木质栏杆的教堂。

5/6. Saint-Jean-Berchmans.
圣约翰贝克曼斯。

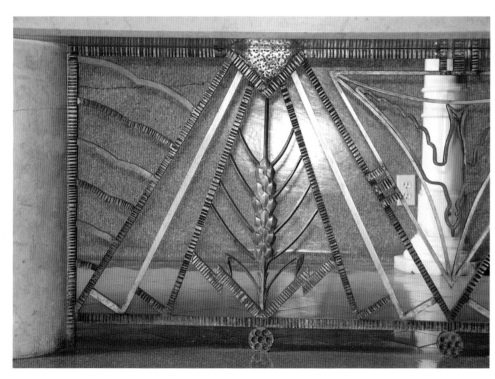

护栏栏杆

ACKNOWLEDGEMENT 鸣谢

We would like to thank everyone for having been involved in the book, especially the following amazing photographers.

本书的编写离不开各位摄影师的帮助，正得益于他们专业而负责的工作态度，才让这本书顺利出版。以下是参与此次出版的摄影师：

©Erik De Graaf, ©Deborah Hewitt, ©Poladamonte, ©Sean Pavone, ©Anthony Villalon, ©Lyn Baxter, ©Rolf52, ©Rorem, ©Umerpk, ©Hudakore, ©Rob Van Esch, ©Jacek Sopotnicki, ©Zhukovsky, ©Antonio Sena, ©Anthony Aneese Totah Jr, ©Benkrut, ©Brandon Seidel, ©Cynthia Farmer, ©David Gilder, ©Izanbar, ©Glenn Nagel, ©Noamfein, ©plasid, ©Ricardo Garza, ©Valentin Armianu, ©William Perry, ©Wollertz, ©Amoklv, ©Anusorn62, ©Charles Sichel-outcalt , ©Crystal Srock, ©Deborah Hewitt, ©Drobm, ©Francesa, ©Halil I. Incim, ©Indiatraveler, ©Jozef Sedmak, ©Kantilal Patel, ©Lazyllama, ©Littleny, ©Littleny, ©Matthew Ragen, ©Margojh, ©Michael Ansell, ©Papasexy, ©Paul Benefield, ©Petr Jilek, ©Popa Sorin, ©Radist, ©Rido, ©Rixie, ©Ruslan Kudrin, ©Steve Allen, ©Swoodie, ©Tanja Rosso, ©William Perry, ©Amichaelbrown, ©Charles Sichel-outcalt, ©Constantin Opris, ©Davidrh, ©Linwendy, ©Mary Lane, ©Mazikab, ©Nopphadon, ©Onion, ©Silva Krajnc, ©Teodora8, ©Vladimir Fedoroff, ©Watthano, ©Zubarciuc Dumitru, ©Andrea Rossi, ©Baloncici, ©Chinkopa, ©Edward Fielding, ©Europhotos, ©Flynt, ©Radka1, ©Fotobym, ©Hywit Dimyadi, ©Igor Aronov, ©Inna Felker, ©Julia Fikse, ©Santos-diez, ©Teodora8, ©Tupungato, ©Glenn Nagel, ©Ivan Vander Biesen, ©Darryl Brooksm, ©Erik De Graaf, ©Mahroch, ©Jacek Sopotnicki, ©Checco, ©Emily Lee, ©Marcio Silva, ©Jorg Hackemann, ©Alex Zarubin, ©Ifeelstock, ©Aleksey Solodov, ©Scaliger, ©Vovez, ©Álvaro Germán Vilela, ©Mario Savoia, ©Mihai-bogdan Lazar, ©Joseph, ©Adam Jones, ©Bill Bradford, ©chicagogeek, ©Hitchster, ©jose luis gil, ©Noah Jeppson, ©Phillip, ©Steve Cadman, ©Terry Robinson, ©Alex, ©arctic_whirlwind, ©Eric Golub, ©Graeme Maclean, ©Jeremy Thompson, ©john babis, ©TANAKA Juuyoh (田中十洋), ©kennymatic, ©taberandrew, ©Rev Stan, ©Simon Blackley, ©William Warby, ©Eduardo, ©MargaretNapier on Dreamstime.com and ©Wonderlane, ©Son of Groucho, ©Neal Jennings, ©Fran Babcock, ©Catherine Read, ©Elliott Brown, ©Bernal Saborio, ©Jane Nearing, ©Daniel X. O'Neil, ©William Warby, ©Pablo, ©traveljunction, ©Miguel Discart, ©Laurie Avocado, ©Rex Brown, ©Alan, ©Chris, ©Chris Sampson, ©Ed Uthman, ©Fabio Achilli, ©Francisco Anzola, ©Francisco Gonzalez, ©Hrag Vartanian, ©Jorge Láscar, ©Ken Lund, ©Nels Olsen, ©poppet with a camera, ©Rain0975, ©Sean Davis, ©Simon Blackley, ©Tim Adam, ©Tony Hisgett, ©William Murphy, ©Haydn Blackey, ©anne beaumont, ©Billie Ward, ©Brad Clinesmith, ©Brian Lamb, ©Carlo Mirante, ©Charlie Dave, ©Claudia Brauer, ©David Brossard, ©David Hilowitz, ©denisbin, ©Eden, Janine and Jim, ©Elliott Brown, ©Franklin Heijnen, ©Jason Eppink, ©Jeppestown, ©Jose Luis Cernadas Iglesias, ©Karen Green, ©Matt Brown, ©no rain corp, ©Phillip Pessar, ©Roman Boed, ©Ron Jones, ©Ruth Hartnup, ©Ryan Dickey, ©Sandra Cohen-Rose and Colin Rose, ©Sean MacEntee, ©shrinkin'violet, ©So_P, ©Spencer Means, ©stephenrwalli, ©Tom Hilton, ©Tom Page, ©William Murphy, ©jcw1967, ©Chris Waits, ©Marcel Oosterwijk, ©Alex Liivet, ©Mark Hogan, ©Andrew Moore, ©elPadawan, ©Pedro Ribeiro Simes, ©Kent Wang, ©denisbin, ©romana klee, ©Love Art Nouveau, ©Jeffrey Zeldman, ©Jason V, ©Chicago Architecture Today ©David Veksler, ©Harvey Barrison, ©Gary Stevens on Flickr.